ROBOTICS

ROBOTICS

APPIN KNOWLEDGE SOLUTIONS

INFINITY SCIENCE PRESS LLC
Hingham, Massachusetts
New Delhi

Infinity Science Press LLC
11 Leavitt Street
Hingham, MA 02043
Tel. 877-266-5796 (toll free)
Fax 781-740-1677
info@infinitysciencepress.com
www.infinitysciencepress.com

This book is printed on acid-free paper.

Appin Knowledge Solutions. *Robotics*.
ISBN: 978-1-934015-02-5

The publisher recognizes and respects all marks used by companies, manufacturers, and developers as a means to distinguish their products. All brand names and product names mentioned in this book are trademarks or service marks of their respective companies. Any omission or misuse (of any kind) of service marks or trademarks, etc. is not an attempt to infringe on the property of others.

Library of Congress Cataloging-in-Publication Data

Robotics / Appin Knowledge Solutions.
        p. cm.
Includes index.
  ISBN 978-1-934015-02-5 (hardcover with cd-rom : alk. paper)
  1. Robotics–Handbooks, manuals, etc. 2. Robots, Industrial–Handbooks, manuals, etc. I. Appin Knowledge Solutions (Firm)
  TJ211.R555 2007
  629.8'92--dc22
                              2007010554

7 8 9 4 3 2 1

Our titles are available for adoption, license, or bulk purchase by institutions, corporations, etc. For additional information, please contact the Customer Service Dept. at 877-266-5796 (toll free).

Requests for replacement of a defective CD-ROM must be accompanied by the original disc, your mailing address, telephone number, date of purchase and purchase price. Please state the nature of the problem, and send the information to Infinity Science Press, 11 Leavitt Street, Hingham, MA 02043.

The sole obligation of Infinity Science Press to the purchaser is to replace the disc, based on defective materials or faulty workmanship, but not based on the operation or functionality of the product.

About the Authors

Appin Knowledge Solutions, with its Asia Pacific Headquarters in New Delhi, is an affiliate of the Appin Group of Companies based in Austin, Texas. Its many businesses span software development, consulting, corporate training programs, and empowerment seminars & products. *Appin Knowledge Solutions* is comprised of prominent industry professionals, many from the University of Texas, Austin and the Indian Institute of Technology, Delhi. The company has extensive experience working with Fortune 500 companies including Microsoft, AT&T, General Electric, & IBM. *Appin Knowledge Solutions* aims to bridge the gap between academia and industry by training people worldwide in fields such as nanotechnology and information security through Appin Technology Labs and its distance learning programs.

This book has been co-authored by the technical team at *Appin Knowledge Solutions*. The group is headed by Rajat Khare and includes the following technology professionals: Ashok Sapra, Ishan Gupta, Vipin Kumar, Anuj Khare, Tarun Wig, and Monika Chawla.

Table of Contents

Chapter 1 INTRODUCTION

In This Chapter

· Introduction to Robotics
· History of Robotics
· Current Research in Robotics around the World
· Classification of Robotics
· An Overview of the Book

1.1 INTRODUCTION TO ROBOTICS

Recently there has been a lot of discussion about futuristic wars between humans and robots, robots taking over the world and enslaving humans. Movies like *The Terminator, Star Wars*, etc., have propogated these ideas faster than anything else. These movies are beautiful works of fiction and present us with an interesting point of view to speculate. However, the truth is much different but equally as interesting as the fiction. If you look around yourself you will see several machines and gizmos within your surroundings. When you use a simple pair of spectacles, do you become non-living? When an elderly person uses a hearing aid or a physically challenged person uses an artificial leg or arm do they become half machine? Yes, they do. Now we are rapidly moving toward an era where we will have chips embedded inside our bodies. Chips will communicate with our biological sensors and will help us in performing several activities more efficiently. An artificial retina is almost at the final stages of its development. Now we are thinking in terms of nanobots helping us to strengthen our immune systems. Now we are already on the verge of becoming half machine. Chips will be implanted inside our bodies imparting telescopic and microscopic abilities in our eyes. Cell phones will be permanently placed inside the ear. We will communicate with different devices not through a control panel or keyboard; rather these devices will receive commands from the brain directly. The next level of development will be the part of the brain being replaced by chips, which will impart more capability to the brain. You may ask, do we need all these? The answer is that the biological evolution has already become obsolete. It is unable to keep pace with the rate at which humans are growing. Many of our primary intuitions, such as mating behavior, are still millions of years old. Evolution happens only after millions of years. But humans have built the entire civilization in only 10,000 years. And now the rate of growth has become exponential. Now we need to replace our brain's decision-making software with faster/better ones. So, where are we heading? Yes, we are slowly becoming robots. Robots are not our competitors on this planet. They are our successors. Robots are the next level in evolution; rather we can call it **robolution**. We will begin our journey with a brief history of robotics.

1.2   HISTORY OF ROBOTICS

Our fascination with robots began more than 100 years ago. Looking back, it's easy to get confused about what is and is not a robot. Robotics' history is tied to so many other technological advances that today seem so trivial we don't even

FIGURE 1.1

FIGURE 1.2

think of them as robots. How did a remote-controlled boat lead to autonomous metal puppies?

Slaves of Steel

The first person to use the word robot wasn't a scientist, but a playwright. Czecho-slovakian writer Karel Capek first used the word robot in his satirical play, *R.U.R.* (Rossum's Universal Robots). Taken from the Czech word for forced labor, the word was used to describe electronic servants who turn on their masters when given emotions. This was only the beginning of the bad-mouthing robots would receive for the next couple of decades. Many people feared that machines would resent their role as slaves or use their steely strength to overthrow humanity.

Wartime Inventions

World War II was a catalyst in the development of two important robot components i.e., artificial sensing and autonomous control. Radar was essential for

FIGURE 1.3

tracking the enemy. The U.S. military also created autocontrol systems for mine detectors that would sit in front of a tank as it crossed enemy lines. If a mine was detected, the control system would automatically stop the tank before it reached the mine. The Germans developed guided robotic bombs that were capable of correcting their trajectory.

Calculators and Computers

Mathematician Charles Babbage dreamed up the idea for an "Analytical Engine" in the 1830s, but he was never able to build his device. It would take another 100 years before John Atanassoff would build the world's first digital computer. In 1946 the University of Pennsylvania completed the ENIAC (Electronic Numerical Integrator and Calculator), a massive machine made up of thousands of vacuum tubes. But these devices could only handle numbers. The UNIVAC I (Universal Automatic Computer) would be the first device to deal with letters.

A Robot in Every Pot

For robotics, the '40s and '50s were full of over-the-top ideas. The invention of the transistor in 1948 increased the rate of electronic growth and the possibilities seemed endless. Ten years later, the creation of silicon microchips reinforced that growth. The Westinghouse robot Elecktro showed how far science and imagination could go. The seven-foot robot could smoke and play the piano. Ads from the era suggested that every household would soon have a robot.

FIGURE 1.4

FIGURE 1.5

Industrial-strength Arms

As the demand for cars grew, manufacturers looked for new ways to increase the efficiency of the assembly line through telecherics. This new field focused on robots that mimicked the operator's movements from a distance. In 1961 General Motors installed the applied telecherics system on their assembly line. The one-armed robot unloaded die casts, cooled components, and delivered them to a trim press. In 1978 the PUMA (Programmable Universal Machine for Assembly) was introduced and quickly became the standard for commercial telecherics.

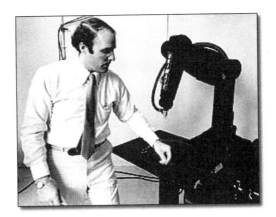

FIGURE 1.6

FIGURE 1.7

Early Personal Robots

With the rise of the personal computer came the personal robot craze of the early '80s. The popularity of *Star Wars* didn't hurt either. The first personal robots looked like R2D2. The RB5X and the HERO 1 robots were both designed as education tools for learning about computers. The HERO 1 featured light, sound, and sonar sensors, a rotating head and, for its time, a powerful microprocessor. But the robots had a lighter side, too. In demo mode, HERO 1 would sing. The RB5X even attempted to vacuum, but had problems with obstacles.

Arms in Space

Once earthlings traveled to space, they wanted to build things there. One of NASA's essential construction tools is the Canadarm. First deployed in 1981

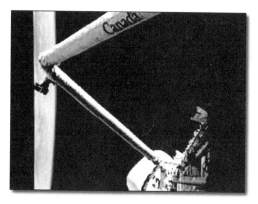

FIGURE 1.8

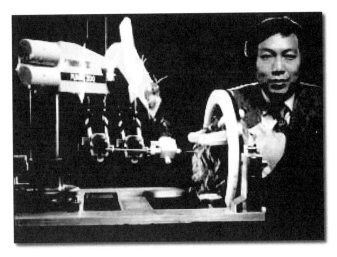

FIGURE 1.9

aboard the Columbia, the Canadarm has gone on to deploy and repair satellites, telescopes, and shuttles. Jet Propulsions Laboratories (JPL) in California has been working on several other devices for space construction since the late eighties. The Ranger Neutral Buoyancy Vehicle's many manipulators are tested in a large pool of water to simulate outer space.

Surgical Tools

While robots haven't replaced doctors, they are performing many surgical tasks. In 1985 Dr. Yik San Kwoh invented the robot-software interface used in the first robot-aided surgery, a stereotactic procedure. The surgery involves a small probe that travels into the skull. A CT scanner is used to give a 3D picture of the brain, so that the robot can plot the best path to the tumor. The PUMA robots are commonly used to learn the difference between healthy and diseased tissue, using tofu for practice.

The Honda Humanoid

The team who created the Honda Humanoid robot took a lesson from our own bodies to build this two-legged robot. When they began in 1986, the idea was to create an intelligent robot that could get around in a human world, complete with stairs, carpeting, and other tough terrain. Getting a single robot mobile in a variety of environments had always been a challenge. But by studying feet and legs, the Honda team created a robot capable of climbing stairs, kicking a ball, pushing a cart, or tightening a screw.

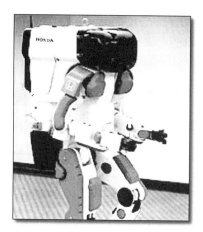

FIGURE 1.10

Hazardous Duties

As scientific knowledge grew so did the level of questioning. And, as with space exploration, finding the answers could be dangerous. In 1994 the CMU Field Robotics Center sent Dante II, a tethered walking robot to explore Mt. Spurr in Alaska. Dante II aids in the dangerous recovery of volcanic gases and samples. These robotic arms with wheels (a.k.a. mobile applied telecherics) saved countless lives defusing bombs and investigating nuclear accident sites. The range of self-control, or autonomy, on these robots varies.

Solar-powered Insects

Some robots mimic humans, while others resemble lower life forms. Mark Tilden's BEAM robots look and act like big bugs. The name BEAM is an acronym

FIGURE 1.11

FIGURE 1.12

for Tilden's philosophy: biology, electronics, aesthetics, and mechanics. Tilden builds simple robots out of discrete components and shies away from the integrated circuits most other robots use for intelligence. Started in the early 1990s, the idea was to create inexpensive, solar-powered robots ideal for dangerous missions such as landmine detection.

A Range of Rovers

By the 1990s NASA was looking for something to regain the public's enthusiasm for the space program. The answer was rovers. The first of these small, semiautonomous robot platforms to be launched into space was the Sojourner, sent to Mars in 1996. Its mission involved testing soil composition, wind speed, and water vapor quantities. The problem was that it could only travel

FIGURE 1.13

FIGURE 1.14

short distances. NASA went back to work. In 2004, twin robot rovers caught the public's imagination again, sending back amazing images in journeys of kilometers, not meters.

Entertaining Pets

In the late '90s there was a return to consumer-oriented robots. The proliferation of the Internet also allowed a wider audience to get excited about robotics, controlling small rovers via the Web or buying kits online. One of the real robotic wonders of the late '90s was AIBO the robotic dog, made by Sony Corp. Using his sensor array, AIBO can autonomously navigate a room and play ball. Even with a price tag of over $2,000, it took less than four days for AIBO to sell out online. Other "pet robots" followed AIBO, but the challenge of keeping the pet smart and the price low remains.

1.3   CURRENT RESEARCH IN ROBOTICS AROUND THE WORLD

According to MSN Learning & Research, 700,000 robots were in the industrial world in 1995 and over 500,000 were used in Japan, about 120,000 in Western Europe, and 60,000 in the United States– and many were doing tasks that are dangerous or unpleasant for humans. Some of the hazardous jobs are handling material such as blood or urine samples, searching buildings for fugitives, and deep water searches, and even some jobs that are repetitive—and these can run 24 hours a day without getting tired. General Motors Corporation uses these robots for spot welding, painting, machine loading, parts transfer, and assembly.

Assembly line robots are the fastest growing because of higher precision and lower cost for labor. Basically a robot consists of:

- A mechanical device, such as a wheeled platform, arm, or other construction, capable of interacting with its environment.
- Sensors on or around the device that are able to sense the environment and give useful feedback to the device.
- Systems that process sensory input in the context of the device's current situation and instruct the device to perform actions in response to the situation.

In the manufacturing field, robot development has focused on engineering robotic arms that perform manufacturing processes. In the space industry, robotics focuses on highly specialized, one-of-kind planetary rovers. Unlike a highly automated manufacturing plant, a planetary rover operating on the dark side of

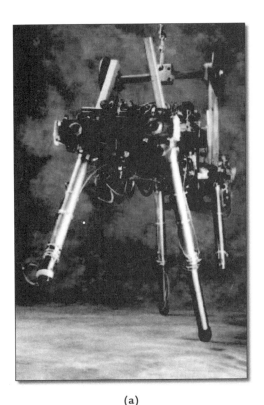

(a)   (b)

FIGURE 1.15   **The older robots of the MIT leg Lab. (a) Quadruped demonstrated that two-legged running algorithms could be generalized to allow four-legged running, including the trot, pace, and bound. (b) The 3D biped hops, runs, and performs tucked somersaults.**

the moon without radio communication might run into unexpected situations. At a minimum, a planetary rover must have some source of sensory input, some way of interpreting that input, and a way of modifying its actions to respond to a changing world. Furthermore, the need to sense and adapt to a partially unknown environment requires intelligence (in other words, artificial intelligence). From military technology and space exploration to the health industry and commerce, the advantages of using robots have been realized to the point that they are becoming a part of our collective experience and everyday lives.

Several universities and research organizations around the world are engaged in active research in various fields of robotics. Some of the leading research organizations are MIT (Massachusetts Institute of Technology), JPL (Jet Propulsion Lab., NASA), CMU (Carnegie Mellon University), and Stanford University.

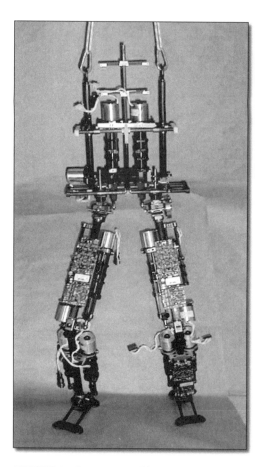 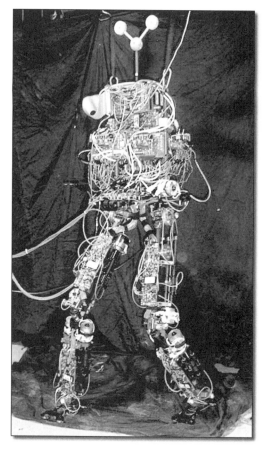

FIGURE 1.16   M2, a 3D bipedal walking robot that is currently being developed in the MIT Leg Laboratory.

These and many other organizations are involved in various fields of robotics. These fields of robotics can be broadly categorized as:

- Robotic Manipulator
- Wheeled Mobile Robots
- Legged Robots
- Underwater Robots
- Flying Robots
- Robot Vision
- Artificial Intelligence
- Industrial Automation

The Leg Lab at MIT is dedicated to studying legged locomotion and building dynamic legged robots. They are specialists in exploring the roles of balance and dynamic control. They are simulating and building creatures which walk, run, and hop like their biological counterparts. The preceeding pictures show a few of their research robots.

FIGURE 1.17   **A JPL space exploration robot.**

M2 is a 3D bipedal walking robot that is currently being developed in the MIT Leg Laboratory. The robot has 12 active degrees of freedom: 3 in each hip, 1 in each knee, and 2 in each ankle. It will be used to investigate:

- Various walking algorithms.
- Motion description and control techniques, particularly Virtual Model Control.
- Force control actuation techniques, particularly Series Elastic Actuation.
- Automatic learning techniques.

Jet Propulsion Laboratory is NASA's lead center for creating robotic spacecraft and rovers. Robots can literally go where no person has gone before, to other planets where the environments are not suitable for humans until we have studied them in much greater detail. The robots and spacecraft we build are our eyes and ears on these distant planets. The preceeding is a picture of a robot that is being developed at JPL.

Carnegie Mellon University is another center that is involved in active research of robotics. There are several robots that are being researched

FIGURE 1.18   Rover1 is a highly autonomous, programmable robot at CMU.

at The Robotics Institute, CMU. One of these robots is Rover 1. One of the goals in designing the rover was to create a robot that could autonomously navigate in the dynamic environment of the home. It uses a visual navigation system dependent on static landmarks. The rover can also climb stairs.

Another project in The Robotics Institute, CMU is Gyrover. Gyrover is a single-wheel robot that is stabilized and steered by means of an internal, mechanical gyroscope. Gyrover can stand and turn in place, move deliberately at low speed, climb moderate grades, and move stably on rough terrain at high speeds. It has a relatively large rolling diameter, which facilitates motion over rough terrain; a single track and narrow profile for obstacle avoidance; and is completely enclosed for protection from the environment.

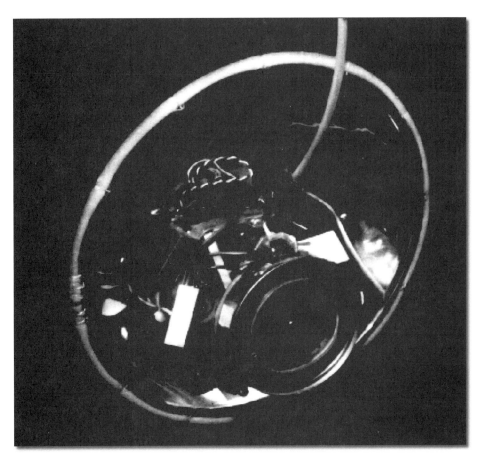

FIGURE 1.19   **Gyrover I, a single-wheel robot that is stabilized and steered by means of an internal, mechanical gyroscope.**

1.4   CLASSIFICATION OF ROBOTICS

As mentioned, robotics can be classified into the following:

- Robotic Manipulator
- Wheeled Mobile Robots (WMR)
- Legged Robots
- Underwater Robots and Flying Robots
- Robot Vision
- Artificial Intelligence
- Industrial Automation

1.4.1   Robotic Arms

Robotic arms have become useful and economical tools in manufacturing, medicine, and other industries.

1.4.2   Wheeled Mobile Robots

Wheeled mobile robots perform many tasks in industry and in the military.

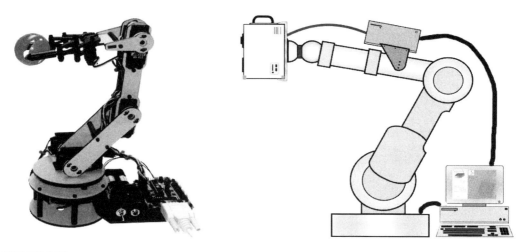

FIGURE 1.20

FIGURE 1.21

1.4.3   Legged Robots

Locomotion on the ground can be realized with three different basic mechanisms:

(i)   Slider,
(ii)   Liver, and
(iii)   Wheel or track.

Out of the above three mechanisms, the first two are walking mechanisms, and in these cases the robot moves on legs. So many robots have been designed that follow the walking mechanism. Walking mechanisms have their own advantages and they become more reasonable when moving on soft, uneven terrains.

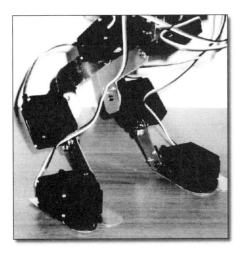 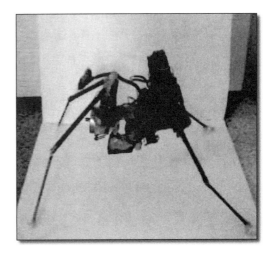

FIGURE 1.22

The benefits that can be obtained with a legged robot are:

- Better mobility
- Better stability on the platform
- Better energy efficiency
- Smaller impact on the ground

When choosing the mechanism for locomotion of a robot, one needs to keep his eyes on the following factors:

- Terrain on which the robot mainly moves
- Operational flexibility needed when working
- Power and/or energy efficiency requirements
- Payload capacity requirements
- Stability
- Impact on the environment

In walking robots, the balance of the body is of prime importance and it becomes even more important if it is a two-legged robot. So the control system used in such robots should be used wisely. A motion control system should control the motion of the body so that leg movements automatically generate the desired body movements.

A control system also needs to control gait i.e., the sequence of supporting leg configurations and foot placement (motion of the nonsupporting legs) to find the next foothold. While walking, the movement of the body which rests on the supporting legs should be considered and properly controlled.

Gait, which determines the sequence of supporting leg configurations during movement, is divided into two classes:

(i) **Periodic gaits:** They repeat the same sequence of supporting leg configurations.
(ii) **Nonperiodic or free gaits:** They do not have any periodicity in their gait pattern.

The number of different gaits depends on the number of legs.

1.4.4 Underwater Robots

Camera-equipped underwater robots serve many purposes including tracking of fish and searching for sunken ships.

FIGURE 1.23

1.4.5 Flying Robots

Flying robots have been used effectively in military maneuvers, and often mimic the movements of insects.

1.4.6 Robot Vision

Vision-based Sensors

Vision is our most powerful sense. It provides us with an enormous amount of information about the environment and enables rich, intelligent interaction in dynamic environments. It is therefore not surprising that a great deal of effort has been devoted to providing machines with sensors that mimic the capabilities of the human vision system. The first step in this process is the creation of the sensing devices that capture the same raw information light that the human vision

FIGURE 1.24

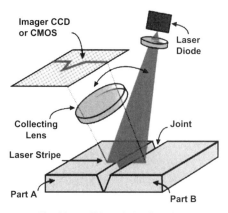

Fig. 1 Laser Triangulation Principle

FIGURE 1.25

system uses. The two current technologies for creating vision sensors are CCD and CMOS. These sensors have specific limitations in performance when compared to the human eye. The vision-based sensors are discussed in detail in Chapter 6.

1.4.7   Artificial Intelligence

Artificial Intelligence (AI) is a branch of computer science and engineering that deals with intelligent behavior, learning, and adaptation in machines. Research in AI is concerned with producing machines to automate tasks requiring intelligent behavior. Examples include control, planning and scheduling, the ability to answer diagnostic and consumer questions, handwriting, speech, and facial recognition. As such, it has become an engineering discipline, focused on providing solutions to real-life problems, software applications, traditional strategy games like computer chess, and other video games.

Schools of Thought

AI divides roughly into two schools of thought: Conventional AI and Computational Intelligence (CI). Conventional AI mostly involves methods now classified as machine learning, characterized by formalism and statistical analysis. This is also known as symbolic AI, logical AI, neat AI, and Good Old-Fashioned Artificial Intelligence (GOFAI). Methods include:

■   **Expert systems:** apply reasoning capabilities to reach a conclusion. An expert system can process large amounts of known information and provide conclusions based on them.

FIGURE 1.26

- Case-based reasoning.
- Bayesian networks.
- **Behavior-based AI:** a modular method of building AI systems by hand.

Computational Intelligence involves iterative development or learning (e.g., parameter tuning in connectionist systems). Learning is based on empirical data and is associated with nonsymbolic AI, scruffy AI, and soft computing. Methods mainly include:

- **Neural networks:** systems with very strong pattern recognition capabilities.
- **Fuzzy systems:** techniques for reasoning under uncertainty, have been widely used in modern industrial and consumer product control systems.
- **Evolutionary computation:** applies biologically inspired concepts such as populations, mutation, and survival of the fittest to generate increasingly

better solutions to the problem. These methods most notably divide into evolutionary algorithms (e.g., genetic algorithms) and swarm intelligence (e.g., ant algorithms).

With hybrid intelligent systems attempts are made to combine these two groups. Expert inference rules can be generated through neural network or production rules from statistical learning such as in ACT-R. It is thought that the human brain uses multiple techniques to both formulate and cross-check results. Thus, integration is seen as promising and perhaps necessary for true AI.

1.4.8   Industrial Automation

Automation, which in Greek means self-dictated, is the use of control systems, such as computers, to control industrial machinery and processes, replacing human operators. In the scope of industrialization, it is a step beyond mechanization. Whereas mechanization provided human operators with machinery to assist them with the physical requirements of work, automation greatly reduces the need for human sensory and mental requirements as well.

FIGURE 1.27

1.5   AN OVERVIEW OF THE BOOK

This book includes different aspects of a robot in modules. It also explores the different fields of robotics. Chapter 2 covers theory of machines and mechanisms, introduction to gears and gear trains, kinematics analysis, and synthesis of mechanisms. Section 2.6 covers a practical guide to using various mechanisms in robotic projects.

Chapter 3 covers the basic introduction to electronics. It lays more stress on the issues related to practical electronic circuit design without going into much of the details of the theory. The chapter covers some fundamentals of sensors and microcontrollers. Section 3.7 covers a practical guide to use Embedded C programming for an 8051 microcontroller. It also covers how to use the parallel port and the serial port of the computer to control a few devices using common programming platforms like C++ and VB. Section 3.8 covers a basic introduction to geared DC motors, stepper motors, and servo motors and practical circuits to interface them with digital systems. Section 3.9 covers tips to use some common things found in the neighborhood in projects.

Chapter 4 goes into the details of wheeled mobile robots, their kinematics, mathematical modeling, and control. Section 4.5 covers the simulation of wheeled mobile robots using ODE23 of MATLAB. A few examples will be presented. The simulation examples are also included on the CD-ROM. Section 4.6 covers the step-by-step construction of the hardware and software of an all-purpose practical research WMR.

Chapter 5 covers the kinematics of robotic manipulators. The topics that this chapter covers are, mapping of frames, forward kinematics, and inverse kinematics. Chapter 5.5 includes the guide to make the hardware and software of a two-link arm and a three-link robotic arm.

Chapter 6 talks about sensors that can be used in robots. Various sensors such as digital encoders, infrared sensors, radio frequency sensors, sonar, active beacons, digital compasses, acceleretometers, gyroscopes, laser rangefinders, etc., will be discussed in this chapter. Section 6.12 includes two practical examples of making sensors and interfacing them with digital circuits.

Chapter 7 covers some basic fundamentals about legged robots. It discusses the issues of static and dynamic balance, inverse pendulum model and the kinematics of leg design. The chapter includes a discussion about the gaits of various legged animals found in nature. Section 7.6 covers the dynamic considerations of leg design such as leg lengths and speed of travel, etc.

There is far more to learn about a cross-disciplinary field such as robotics than can be contained in this single book. We hope that this will be enough to place the reader in a comfortable position in the dynamic and challenging field of robotics.

Chapter **2** **BASIC MECHANICS**

In This Chapter

- Introduction to Theory of Machines and Mechanisms
- Some Popular Mechanisms
- Gear and Gear Trains
- Synthesis of Mechanisms
- Kinematic Analysis of Mechanisms
- A Practical Guide to Use Various Mechanisms

2.1 INTRODUCTION TO THEORY OF MACHINES AND MECHANISMS

A mechanism is a device that transforms motion to some desirable pattern and typically develops very low forces and transmits little power. A machine typically contains mechanisms that are designed to provide significant forces and transmit significant power. Some examples of typical mechanisms are a stapler, a door lock, car window wiper, etc. Some examples of machines that possess motions similar to the mechanisms above are an automobile engine, a crane, and a robot. There is no clear line of difference between mechanisms and machines. They differ in degree rather than definition.

If a mechanism involves light forces and is run at slow speeds, it can some-times be strictly treated as a kinematic device; that is, it can be analyzed kine-matically without regard to forces. Machines (and mechanisms running at higher speeds), on the other hand, must be first treated as mechanisms. A kinematic analysis of their velocities and accelerations must be done and then they must be treated as dynamic systems in which their static and dynamic forces due to accelerations are analyzed using the principles of kinetics. Most of the applica-tions in robotics involve motions at lower speeds and low or moderate forces are involved. So we will restrict our discussion only to the kinematics of mechanisms in this chapter. However, there are certain instances where the study of the dy-namics becomes very essential in robotics. A discussion of those instances is beyond the scope of this book.

2.2   SOME POPULAR MECHANISMS

2.2.1   Four-bar Mechanism

In the range of planar mechanisms, the simplest group of lower pair mechanisms is four-bar linkages. A four-bar linkage comprises four bar-shaped links and four turning pairs as shown in Figure 2.1.

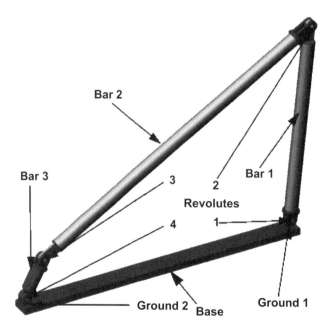

FIGURE 2.1   Four-bar linkage.

The link opposite the frame is called the coupler link, and the links, which are hinged to the frame, are called side links. A link, which is free to rotate through 360 degrees with respect to a second link, will be said to revolve relative to the second link (not necessarily a frame). If it is possible for all four bars to become simultaneously aligned, such a state is called a change point.

Some important concepts in link mechanisms are:

1. **Crank:** A side link, which revolves relative to the frame, is called a crank.
2. **Rocker:** Any link that does not revolve is called a rocker.
3. **Crank-rocker mechanism:** In a four-bar linkage, if the shorter side link revolves and the other one rocks (i.e., oscillates), it is called a crank-rocker mechanism.
4. **Double-crank mechanism:** In a four-bar linkage, if both of the side links revolve, it is called a double-crank mechanism.
5. **Double-rocker mechanism:** In a four-bar linkage, if both of the side links rock, it is called a double-rocker mechanism.

Before classifying four-bar linkages, we need to introduce some basic nomenclature. In a four-bar linkage, we refer to the line segment between hinges on a given link as a **bar** where:

- s = length of the shortest bar
- l = length of the longest bar
- p, q = lengths of the intermediate bars

Grashof's theorem states that a four-bar mechanism has at least one revolving link if

$$s + l <= p + q \tag{2.1}$$

and all three mobile links will rock if

$$s + l > p + q. \tag{2.2}$$

All four-bar mechanisms fall into one of the four categories listed in Table 2.1.

From Table 2.1 we can see that for a mechanism to have a crank, the sum of the length of its shortest and longest links must be less than or equal to the sum of the length of the other two links. However, this condition is necessary but not sufficient. Mechanisms satisfying this condition fall into the following three categories:

TABLE 2.1 Classification of Four-bar Mechanisms			
Case	l + s vers. p + q	Shortest Bar	Type
1	<		
Frame	Double-crank		
2	<		
Side	Rocker-crank		
3	<		
Coupler	Double-rocker		
4	=	Any	Change point
5	>	Any	Double-rocker

1. When the shortest link is a side link, the mechanism is a crank-rocker mechanism. The shortest link is the crank in the mechanism.
2. When the shortest link is the frame of the mechanism, the mechanism is a double-crank mechanism.
3. When the shortest link is the coupler link, the mechanism is a double-rocker mechanism.

2.2.2   Slider-crank Mechanism

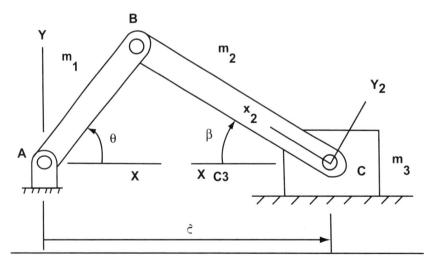

FIGURE 2.2   Parameters of slider-crank mechanisms.

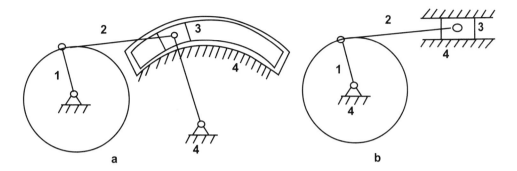

FIGURE 2.3   **Slider-crank mechanism.**

The slider-crank mechanism, which has a well-known application in engines, is a special case of the crank-rocker mechanism. Notice that if rocker 3 in Figure 2.3a is very long, it can be replaced by a block sliding in a curved slot or guide as shown. If the length of the rocker is infinite, the guide and block are no longer curved. Rather, they are apparently straight, as shown in Figure 2.3b, and the linkage takes the form of the ordinary slider-crank mechanism.

Inversion of the Slider-crank Mechanism

Inversion is a term used in kinematics for a reversal or interchange of form or function as applied to kinematic chains and mechanisms. For example, taking a different link as the fixed link, the slider-crank mechanism shown in Figure 2.4a can be inverted into the mechanisms shown in Figures 2.4b, c, and d. Different examples can be found in the application of these mechanisms. For example, the mechanism of the pump device in Figure 2.5 is the same as that in Figure 2.4b.

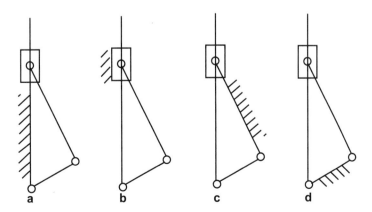

FIGURE 2.4   **Inversions of the crank-slide mechanism.**

Basic Mechanics

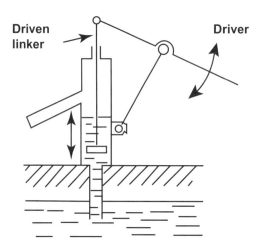

FIGURE 2.5   A pump device.

Keep in mind that the inversion of a mechanism does not change the motions of its links relative to each other but does change their absolute motions.

2.2.3   Rack and Pinion

A 'rack and pinion' gears system looks quite unusual. However, it is still composed of two gears. The 'pinion' is the normal round gear and the 'rack' is straight or flat. The 'rack' has teeth cut in it and they mesh with the teeth of the pinion gear.

FIGURE 2.6

A good example of a 'rack and pinion' gear system can be seen on trains that are designed to travel up steep inclines. The wheels on a train are steel and they have no way of gripping the steel track. Usually the weight of the train is enough to allow the train to travel safely and speed along the track. However, if a train has to go up a steep bank or hill it is likely to slip backward. A 'rack and pinion' system is added to some trains to overcome this problem. A large gear wheel is added to the center of the train and an extra track with teeth, called a 'rack,' is added to the track. As the train approaches a steep hill or slope the gear is lowered to the track and it meshes with the 'rack.' The train does not slip backward but it is pulled up the steep slope.

The railway system in Switzerland is probably the most advanced in the world. The entire system is punctual and modern as a result of financial investment in railway building and locomotive technology. The cities and towns are linked by fast, efficient trains running mainly on a electrified network. On the other hand, the mountain trains rely on a combination of electrified track and modern steam engines. The Swiss have been careful to develop a rail system that is not only extremely efficient but also sensitive to the environment.

On such an incline, using a normal track is not practical as the locomotive would simply slip backward, down the track. The Figure 2.7 below shows the typical 'rack and pinion' track system. The two outer rails are the same as any normal rail track. However, the center track has teeth similar to those seen on a gear wheel (this is called the 'rack'). When the locomotive is required to go up a steep incline a gear wheel (called the 'pinion') is lowered from the locomotive engine. This meshes with the rack and pulls the locomotive and carriages up the steep slope.

FIGURE 2.7   **Track with a rack.**

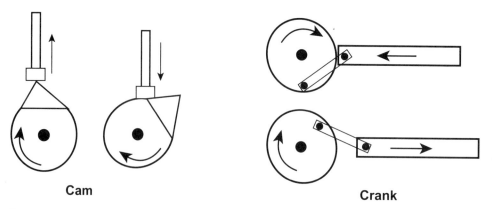

Cam   Crank

FIGURE 2.8 Cams and cranks.

2.2.4   Cams and Cranks

Both cams and cranks are useful when a repetitive motion is desired. Cams make rotary motion a little more interesting by essentially moving the axle off-center. Cams are often used in conjunction with a rod. One end of the rod is held flush against the cam by a spring. As the cam rotates the rod remains stationary until the "bump" of the cam pushes the rod away from the cam's axle.

Cranks convert rotary motion into a piston-like linear motion. The best examples of cranks in action are the drive mechanism for a steam locomotive and the automobile engine crankshaft. In a crank, the wheel rotates about a centered axle, while an arm is attached to the wheel with an off-centered peg. This arm is attached to a rod fixed in a linear path. A crank will cause the rod to move back and forth, and if the rod is pushed back and forth, it will cause the crank to turn. On the other hand, cams can move their rods, but rods cannot move the cams. Cams can be used to create either a linear repetitive motion or a repetitive rotational motion such as the one shown in Figure 2.8.

2.3   GEAR AND GEAR TRAINS

A wheel and axle assembly becomes especially useful when gears and belts are brought into the picture. Gears can be used to change the direction or speed of movement, but changing the speed of rotation inversely affects the force transmitted. A small gear meshed with a larger gear will turn faster, but with less force. There are four basic types of gears: spur gears, 'rack and pinion' gears, bevel gears, and worm gears. Spur gears are probably the type of gear that most people picture when they hear the word. The two wheels are in the same plane (the axles are parallel). With 'rack and pinion' gears there is one wheel and one 'rack', a flat toothed bar that converts the rotary motion into linear motion. Bevel gears are also known as pinion and crown or pinion and ring gears. In bevel gears, two wheels intermesh at an angle changing the direction of rotation (the axles are not parallel); the speed and force may also be modified, if desired. Worm gears involve one wheel gear (a pinion) and one shaft with a screw thread wrapped around it. Worm gears change the direction of motion as well as the speed and force. Belts work in the same manner as spur gears except that they do not change the direction of motion.

In both gears and belts, the way to alter speed and force is through the size of the two interacting wheels. In any pair, the bigger wheel always rotates more slowly, but with more force. This "tradeoff" between force and speed comes from the difference in the distance between the point of rotation and the axle between the two wheels. On both the big and the small gear, the linear velocity at the point of contact for the wheels is equal. If it was unequal and one gear were spinning faster than the other at the point of contact, then it would rip the teeth right off of the other gear. As the circumference of the larger gear is greater, a point on the outside of the larger gear must cover a greater distance than a point on the smaller gear to complete a revolution. Therefore the smaller gear must complete more revolutions than the larger gear in the same time span. (It's rotating more quickly.) The force applied to the outer surface of each wheel must also be equal otherwise one of them would be accelerating more rapidly than the other and again the teeth of the other wheel would break. The forces of interest, however, are not the forces being applied to the outer surfaces of the wheels, but rather the forces on the axles. Returning to the concept of levers, we know that the distance at which the force is applied affects the force yielded, and a wheel and axle works like a lever. Equal forces are being applied to each wheel, but on the larger wheel that force is being applied over a greater distance. Thus for the larger wheel the force on the axle is greater than the force on the axle for the smaller wheel.

Gear Train

A gear train consists of one or more gear sets intended to give a specific velocity ratio, or change direction of motion. Gear and gear train types can be grouped based on their application and tooth geometry.

2.3.1   Spur Gears

The most common type of gear wheel, spur gears, are flat and have teeth projecting radially and in the plane of the wheel (see Figure 2.9). The teeth of these "straight-cut gears" are cut so that the leading edges are parallel to the line of the

TABLE 2.2 Gear Types Grouped According to Shaft Arrangement			
Parallel Axes	**Intersecting Axes**	**Nonintersecting (Nonparallel) Axes**	**Rotary to Translation**
Spur gears	Bevel gears:	Hypoid gears	Rack and pinion
Helical gears	Straight bevel	Crossed helical gears	
Herringbone or double helical gears	Zerol bevel Spiral bevel	Worm gears	

axis of rotation. These gears can only mesh correctly if they are fitted to parallel axles.

Spur gears are inexpensive to manufacture. And they cause no axial thrust between gears. Although they give lower performance, they may be satisfactory in low-speed or simple applications.

2.3.2 Helical Gears

Helical gears offer a refinement over spur gears. The teeth are cut at an angle, allowing for more gradual, hence smoother, meshing between gear wheels, eliminating the whine characteristic of straight-cut gears. A disadvantage of helical gears is a resultant thrust along the axis of the gear, which needs to be accommodated by appropriate thrust bearings, and a greater

FIGURE 2.9 **Spur gear.**

degree of sliding friction between the meshing teeth, often addressed with specific additives in the lubricant. Whereas spur gears are used for low-speed applications and those situations where noise control is not a problem, the use of helical gears is indicated when the application involves high speeds, large power transmission, or where noise abatement is important. The speed is considered to be high when the pitch line velocity (i.e., circumferential velocity) exceeds 5,000 ft./min. or the rotational speed of the pinion (i.e., smaller gear) exceeds 3,600 rpm.

FIGURE 2.10   **Helical gears.**

2.3.3   Bevel Gears

Where two axles cross at point and engage by means of a pair of conical gears, the gears themselves are referred to as bevel gears. These gears enable a change in the axes of rotation of the respective shafts, commonly 90°. A set of four bevel gears in a square make a differential gear, which can transmit power to two axles spinning at different speeds, such as those on a cornering automobile.

Helical gears can also be designed to allow a ninety-degree rotation of the axis of rotation.

FIGURE 2.11   **Bevel gear in a floodgate.**

2.3.4   Worm and Wheel

The arrangement of gears seen below is called a worm and worm wheel. The worm, which in this example is brown in color, only has one tooth, but it is like a screw thread. The worm wheel, colored yellow, is like a normal gear wheel or spur gear. The worm always drives the worm wheel round, it is never the opposite way as the system tends to lock and jam. The picture on the left is a typical setup for a motor and worm gear system. As the worm revolves, the worm wheel (spur gear) also revolves but the rotary motion is transmitted through a ninety-degree angle.

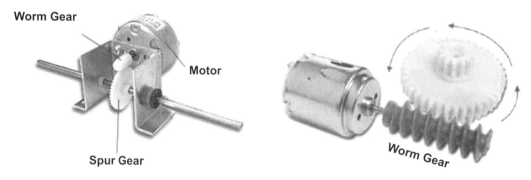

FIGURE 2.12

The worm is always the driving gear. The worm gear can achieve a higher gear ratio than spur gears of a comparable size. Designed properly, a built-in safety feature can be obtained: This gear style will self-lock if power is lost to the drive (worm). It doesn't work if the pinion is powered.

2.3.5   Parallel Axis Gear Trains

Gear trains consist of two or more gears for the purpose of transmitting motion from one axis to another. Ordinary gear trains have axes, relative to the frame, for all gears comprising the train. Figure 2.13a shows a simple ordinary train in which there is only one gear for each axis. In Figure 2.13b a compound ordinary train is seen to be one in which two or more gears may rotate about a single axis.

This is a good example of a 'gear train.' A gear train is usually made up of two or more gears. The driver in this example is gear 'A.' So far you have read about 'driver' gears, 'driven' gears, and gear trains. An 'idler' gear is another important

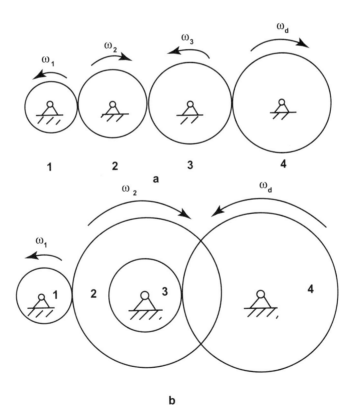

FIGURE 2.13   Ordinary gear trains.

gear. Gear 'A' turns in a counter clockwise direction and also gear 'C' turns in a counter clockwise direction. The 'idler' gear is used so that the rotation of the two important gears is the same.

Velocity Ratio

We know that the velocity ratio of a pair of gears is the inverse proportion of the diameters of their pitch circle, and the diameter of the pitch circle equals the number of teeth divided by the diametric pitch. Also, we know that it is necessary for the mating gears to have the same diametric pitch to satisfy the condition of correct meshing. Thus, we infer that the velocity ratio of a pair of gears is the inverse ratio of their number of teeth.

For the ordinary gear trains in Figure 2.13a, we have:

$$\frac{\omega_1}{\omega_2} = \frac{N_2}{N_1} \quad \frac{\omega_2}{\omega_3} = \frac{N_3}{N_2} \quad \frac{\omega_3}{\omega_4} = \frac{N_4}{N_3}. \tag{2.3}$$

These equations can be combined to give the velocity ratio of the first gear in the train to the last gear:

$$\frac{\omega_1}{\omega_4} = \frac{N_2 N_3 N_4}{N_1 N_2 N_3} = \frac{N_4}{N_1}. \tag{2.4}$$

NOTE

- The tooth numbers in the numerator are those of the driven gears, and the tooth numbers in the denominator belong to the driver gears.
- Gear 2 and 3 both drive and are, in turn, driven. Thus, they are called idler gears. Since their tooth numbers cancel, idler gears do not affect the magnitude of the input-output ratio, but they do change the directions of rotation. Note the directional arrows in Figure 2.13. Idler gears can also constitute a saving of space and money (if gear 1 and 4 mesh directly across a long center distance, their pitch circle will be much larger).
- There are two ways to determine the direction of the rotary direction. The first way is to label arrows for each gear as in Figure 2.13. The second way is to multiple mth power of "-1" to the general velocity ratio. Where m is the number of pairs of external contact gears (internal contact gear pairs do not change the rotary direction). However, the second method cannot be applied to the spatial gear trains.

Thus, it is not difficult to get the velocity ratio of the gear train in Figure 2.13b:

$$\frac{\omega_1}{\omega_4} = (-1)^2 \frac{N_2 N_4}{N_1 N_3}. \tag{2.5}$$

2.4   SYNTHESIS OF MECHANISMS

Most engineering design practice involves a combination of synthesis and analysis. However, one cannot analyze anything until it is synthesized into existence. Many mechanism design problems require the creation of a device with certain motion characteristics. An example could be moving a tool from position A to position B in a particular time interval. There could be endless possibilities. But a common denominator is often the need for a linkage to generate the desired motions. So, we will now explore some simple synthesis techniques to enable you to create potential linkage design solutions for some typical kinematic applications.

2.4.1   Type, Number, and Dimensional Synthesis

Synthesis of a mechanism in most situations cannot be done in a strictly defined manner. Since most real problems have more unknown variables than the number of equations that describes the system's behavior, you cannot simply solve the equations to get a solution. So there are a number of methods available to approach the problem of synthesis. Each method can be approached in a qualitative or quantitative manner. If the number of unknowns is more than or equal to the number of equations, then quantitative synthesis can be employed. Most of the practical design situations actually have a lesser number of equations as compared to the number of variables. Hence, those cases are solved by a qualitative approach.

Type Synthesis refers to the definition of a proper type of mechanism best suited to the problem. It is a qualitative method and it requires some experience and knowledge of the various types of mechanisms that exist and which also may be feasible from a performance and manufacturing standpoint. As an example, assume that the task is to design a device to track the straight line motion of a part on a conveyor belt and spray it with a chemical coating as it passes by. This has to be done at high, constant speed, with good accuracy and repeatability, and it must be reliable. Unless you have had the opportunity to see a wide variety of mechanical equipment, you might not be aware that this task could conceivably be achieved by any of the following devices:

- A straight line linkage
- A cam and follower
- An air cylinder
- A hydraulic cylinder
- A robot
- A solenoid

Each of these solutions, while possible, may not be optimal or even practical. More details need to be known about the problem to make that judgment. The straight-line linkage may prove to be too large and to have undesirable accelerations; the cam and follower will be expensive, though accurate and repeatable. The air cylinder is inexpensive, but noisy and unreliable. The hydraulic cylinder is more expensive as is the robot. The solenoid, while cheap, has high impact loads and high impact velocity. So, you can see that the choice of the device type can have a large effect on the quality of design. A poor design in the type synthesis stage can create insoluble problems later on.

Dimensional Synthesis of a linkage is the determination of the proportions (lengths) of the links necessary to accomplish the desired motion. It can be a form of quantitative synthesis if an enough number of equations is available, but can also be a form of qualitative synthesis. Dimensional synthesis of cams is usually quantitative. However, dimensional synthesis of linkages is usually qualitative. Dimensional synthesis assumes that, through type synthesis, you have already determined that a linkage (or cam) is the most appropriate solution to the problem. We will discuss some analytical and graphical dimensional synthesis of linkages in the following sections of this chapter.

Number Synthesis of a linkage is done to determine the number and order of links and joints necessary to produce motion of a particular DOF. Link order in this context refers to the number of nodes per link, i.e., binary, quaternary, ternary, etc. The value of number synthesis is to allow the exhaustive determination of all possible combinations of links that will yield any chosen DOF. This then equips the designer with a definitive catalog of potential linkages to solve a variety of motion-control problems. Using the number synthesis method to determine all possible link configurations for one DOF motion, we can arrive at the following conclusion. There is only one four-link configuration, two six-link configurations, and five possibilities for an eight-link configuration.

2.4.2   Function Generation, Path Generation, and Motion Generation

Function generation is defined as the correlation of an input motion with an output motion in a mechanism. A function generator is conceptually a black box that delivers some predictable output in response to a known input. Historically, before the advent of electronic computers, mechanical function generators found wide application in artillery rangefinders and shipboard gun-aiming systems, and many other tasks. They are in fact mechanical analog computers. The development of inexpensive digital electronic microcomputers for control systems coupled with the availability of compact servomotors has reduced the demand for these mechanical function generator linkage devices. Many such applications can now be served more economically and efficiently with electromechanical devices. Moreover, the computer-controlled electromechanical function generator is programmable, allowing rapid modification in the function generated as demand changes. The cam-follower system is a form of mechanical function generator, and it is typically capable of higher force and power levels per dollar than electromechanical systems.

Path generation is defined as the control of a point in the plane such that it follows some prescribed path. This is typically achieved with at least four bars, wherein a point in the coupler traces the desired path. No attempt is made in path generation to control the orientation of the link that contains the point of interest. However, it is common for the timing of the arrival of the point at particular locations along the path to be defined. This case is called path generation with prescribed timing and is analogous to function generation in that a particular output function is specified.

Motion generation is defined as the control of a line in the plane such that it assumes some prescribed set of sequential positions. Here the orientation of the link containing the line is important. This is a more general problem than path generation. In fact, path generation is a subset of motion generation. An example of a motion generation problem is the control of the bucket on a bulldozer. The bucket must assume a set of positions to dig, pick up, and dump the excavated earth. Conceptually, the motion of a line, painted on the side of the bucket, must be made to assume the desired positions. A linkage is the usual solution.

2.4.3   Two-position Synthesis

Two-position synthesis is a dimensional synthesis, which can be performed on any mechanism after the mechanism has been finalized in the type synthesis stage. There are several methods to accomplish this task. We will discuss the graphical method for a four-bar linkage. The graphical method is the simplest and quickest method for two-position synthesis. Simplest by no means suggests that it compromises over the quality of solution. The principles used in this graphical synthesis technique are simply those of Euclidean geometry. The following definitions are restated here, since they will be used repeatedly in the following texts.

1. **Crank:** A side link, which revolves relative to the frame, is called a crank.
2. **Rocker:** Any link that does not revolve is called a rocker.
3. **Crank-rocker mechanism:** In a four-bar linkage, if the shorter side link revolves and the other side rocks (i.e., oscillates), it is called a crank-rocker mechanism.

Here we will design a four-bar crank-rocker mechanism to give 45° of rocker rotation with equal time forward and back, from a constant speed motor input.

You have to follow the following steps to design the above mechanism.

1. Draw the output link O_4B in both extreme positions, B_1 and B_2 in any convenient location, such that the desire angle of motion θ_4 is subtended.
2. Draw the chord B_1B_2 and extend it in either direction.
3. Select a convenient point O_2 on line B_1B_2 extended.
4. Bisect line segment B_1B_2, and draw a circle of that radius about O_2.

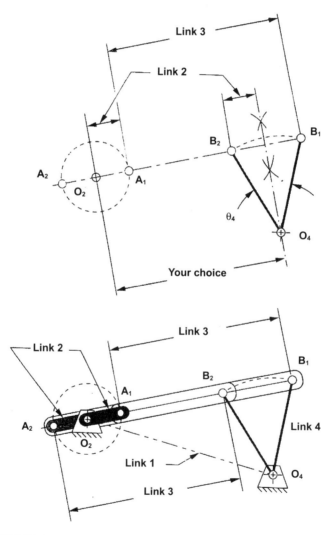

FIGURE 2.14   Two-position function synthesis with rocker output (non-quick-return).

5. Label the two intersections of the circle B_1B_2 and extended A1 and A_2.
6. Measure the length of the coupler as A_1 to B_1 or A_2 to B_2.
7. Measure ground length 1, crank length 2, and rocket length 4.
8. Find the Grashof condition. If non-Grashof, redo steps 3 to 8 with O_2 further from O_4.
9. Make a cardboard model of the linkage and articulate it to check its functions and its transmission angles.
10. You can input the file FO3-04.4br to program FOURBAR to see this example come alive.

Note several things about this synthesis process. We started with the output end of the system, as it was the only aspect defined in the problem statement. We had to make many quite arbitrary decisions and assumptions to proceed because there were many more variables that we could have provided "equations" for. We are frequently forced to make "free choices" of "a convenient angle or length." These free choices are actually definitions of design parameters. A poor choice will lead to a poor design. Thus these are qualitative synthesis approaches and require an iterative process, even for this simple an example. The first solution you reach will probably not be satisfactory, and several attempts (iterations) should be expected to be necessary. As you gain more experience in designing kinematics solutions you will be able to make better choices for these design parameters with less iteration. A simple cardboard model should be made after this stage. You get the most insight into your design's quality for the least effort by making, articulating, and studying the model. Figure 2.14 represents the graphical synthesis method pictorially. It is much easier to understand this method from the pictures.

The problem discussed here is one of the simplest synthesis problems. More complicated synthesis can also be done using the above method. However, there is a detailed analysis of which is beyond the scope of this book. The interested reader can refer to texts of kinematics of mechanisms and theory of machines for a detailed knowledge of the above topic.

The two-position graphical synthesis method is discussed in the following example applied in a practical situation for clearer illustration.

Application of the two-position method to a practical situation

This method uses the principle that a link pinned to the truck will move in a circle. Since we know that any point along a perpendicular bisector is equidistant to the endpoints, it can be deduced that the distance to the endpoints forms the radius of a circle that the link would follow. Note that this method allows you to choose a point along a line, while the next approach constrains you to just two points.

2.4.4 Three-position Synthesis

Three-position synthesis allows the definition of three positions of a line in the plane and will create a four-bar linkage configuration to move it to each of those positions. This is a motion generation problem. The synthesis technique is a

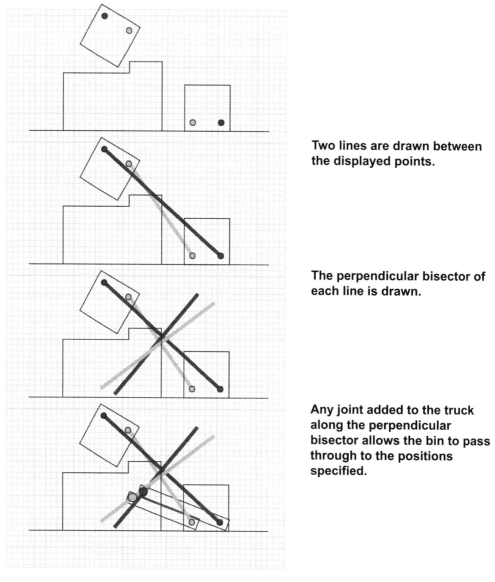

Two lines are drawn between the displayed points.

The perpendicular bisector of each line is drawn.

Any joint added to the truck along the perpendicular bisector allows the bin to pass through to the positions specified.

FIGURE 2.15 Two points (red and gray) are chosen on the bin and plotted on the two desired positions. This can be applied to transformations in a geometry classroom.

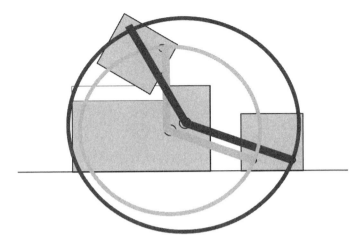

FIGURE 2.16   Alternate constraining approach.

logical extension of the method used in the previous section for two-position synthesis. Compass, protector, and rule are the only tools needed in the graphical method.

Here we will design a four-bar linkage to move the link CD shown from position C_1D_1 to C_2D_2 and then to position C_3D_3. Moving pivots are at C and D.

1. Draw link CD in three-design position C_1D_1, C_2D_2, C_3D_3 in the plane as shown in Figure 2.17.
2. Draw construction lines from point C_1 to C_2 and from C_2 to C_3.
3. Bisect line C_1C_2 and line C_2C_3 and extend their perpendicular bisectors until they intersect. Label their intersection O_2.
4. Repeat steps 2 and 3 for lines D_1D_2 and D_2D_3. Label the intersection O_4.
5. Connect O_2 with C_1 and call it link 2. Connect O_4 with D_1 and call it link 4.
6. Line C_1D_1 is link 3. Line O_2O_4 is link 1.
7. Check the Grashof condition. Note that any Grashof condition is potentially acceptable in this case.
8. Construct a cardboard model and check its function to be sure it can get from initial to final position without encountering any limit (toggle) positions.

Note that while a solution is usually obtained for this case, it is possible that you may not be able to move the linkage continuously from one position to the next without disassembling the links and reassembling them to get them past a limiting position. That will obviously be unsatisfactory. In the particular solution presented in Figure 2.17, note that link 3 and 4 are in toggle at position one, and

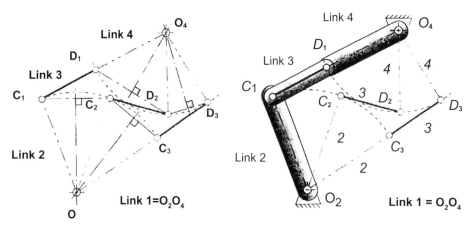

(a) Construction method.   (b) Finished non-Grashof four bar.

FIGURE 2.17

link 2 and 3 are in toggle position at position three. In this case we will have to drive link 3 with a driver dyad, since any attempt to drive either link 2 or link 4 will fail at the toggle positions. No amount of torque applied to link 2 at position C_1 will move link 4 away from point D_1, and driving link 4 will not move link 2 away from position C_3.

The position graphical synthesis method is discussed for a practical situation in the following.

This method uses the same principle as the previous method, but it allows us to prescribe one more position. This is good to account for not hitting the truck cab, but bad because we constrain ourselves to using the two points determined by the intersection of the two mid-normals. Do you think that it would be possible to prescribe 4 positions? Why or why not?

There are many other approaches to synthesize mechanisms. We have discussed the graphical synthesis method in this text. The graphical synthesis method is highly intuitive and depends heavily on the experience and expertise of the person doing the synthesis. However, the graphical synthesis method becomes exceedingly difficult for more than three-position synthesis. For problems involving synthesis of more than three points, the analytical method is applied. The analytical synthesis method is not discussed in this text. The purpose of this topic is to introduce the idea of synthesis. The interested reader can refer to texts of kinematics of mechanisms and theory of machines for a more rigorous study of the above topic.

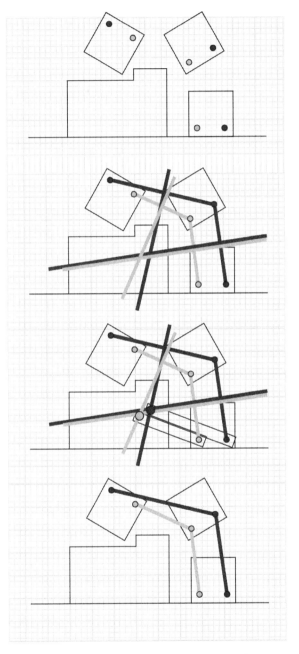

The perpendicular bisectors are drawn on each line.

The intersection of corresponding mid-normals gives the base positions of each of the links (the other being the two points on the coupler link).

Lines are drawn between the displaced points.

FIGURE 2.18 Two points (red and gray) are chosen on the bin and plotted on the three desired positions.

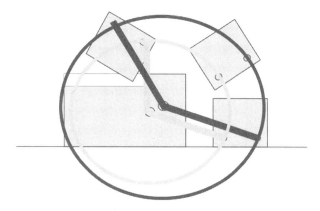

FIGURE 2.19

2.5   KINEMATIC ANALYSIS OF MECHANISMS

After the mechanism has been synthesized, it must be analyzed. A principal goal of kinematic analysis is to determine the accelerations of all the moving parts in the assembly, since dynamic forces are proportional to acceleration. We need to know the dynamic forces in order to calculate the stresses in the components. In order to calculate accelerations, we must first find the positions of all the links or elements in the mechanism for each increment of input motion, and then differentiate the position equations versus time to find velocities, and then to differentiate again to obtain the expression for acceleration.

This can be done in several methods. We could use a graphical approach to determine the position, velocity, and acceleration of the output links for all 180 positions of interest, or we could derive the general equations of motion for any position, differentiate for velocity and acceleration, and then solve these analytic expressions for our 180 (or more) crank rotations. A computer will make this task much easier. If we choose the graphical approach to analysis, we have to do an independent graphical solution for each of the positions of interest. In contrast, once the analytical solution is derived for a particular mechanism, it can be quickly solved (with a computer) for all positions. In this chapter, we will discuss the graphical method and two analytical methods, namely the algebraic method and the complex algebra method, for the position analysis of a few planar mechanisms.

2.5.1   Graphical Position Analysis Method

For any one-DOF linkage, such as a four bar, only one parameter is needed to completely define the positions of all the links. The parameter usually chosen is

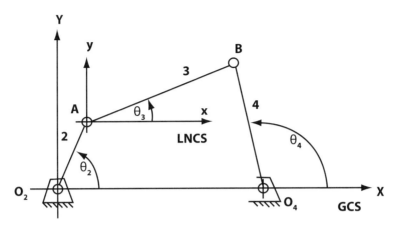

FIGURE 2.20 Measurement of angles in the four-bar linkage.

the angle of the input link. This is shown as θ_2 in Figure 2.20. We want to find θ_3 and θ_4. The link lengths are known. Note that we will consistently number the ground link as 1 and the driver link as 2 in these examples.

The graphical analysis of the problem is trivial and can be done using only high school geometry. If we draw the linkage carefully to scale with rule, compass, and protector in a particular position (given θ_2), then it is only necessary to measure the angles of links 3 and 4 with the protractor. Note that all link angles are measured from a positive x-axis. In Figure 2.18, a local x-y axis system, parallel to the global XY system, has been created at point A to measure θ_3. The accuracy of this graphical solution will be limited by our care and drafting ability and by the crudity of the protractor used. Nevertheless, a very rapid, approximate solution can be found for any one position.

Figure 2.21 shows the construction of the graphical position solution. The four link lengths a, b, c, d and the angle θ_2 of the input link are given. First, the ground link (1) and the input link (2) are drawn to a convenient scale such that they intersect at the origin O_2 of the global XY coordinate system with link 2 placed at the input angle θ_2. Link 1 is drawn along the x-axis for convenience. The compass is set to the scale length of link 3, and an arc of that radius swung about the end of link 2 (point A). Then the compass is set to the scale length of link 4, and a second arc swung about the end of link 1 (point θ_4).

These two arcs will have two intersections at B and B' that define the two solutions to the position problem for a four-bar linkage that can be assembled in two configurations, called circuits, labeled open and crossed in Figure 2.21. Circuits in linkages will be discussed in a later section.

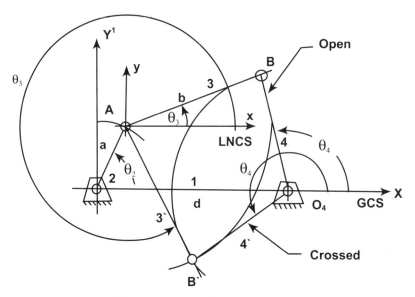

FIGURE 2.21   **Graphical position solution to the open and crossed configurations of the four-bar linkage.**

The angles of link 3 and 4 can be measured with a protractor. One circuit has angles θ_3 and θ_4, the other $\theta_{3'}$ and θ_4'. A graphical solution is only valid for the particular value of input angle used. For each additional position analysis we must completely redraw the linkage. This can become burdensome if we need a complete analysis at every 1- or 2-degree increment of θ_2. In that case we will be better off to derive an analytical solution for θ_3 and θ_4, which can be solved by computer.

2.5.2   Algebraic Position Analysis of Linkages

The same procedure that was used in Figure 2.21 to solve geometrically for the intersections B and B' and angles of links 3 and 4 can be encoded into an algebraic algorithm. The coordinates of point A are found from:

$$A_x = a \cos \theta_2$$

$$A_x = a \sin \theta_2.$$

The coordinates of point B are found using the equations of circles about A and O_4

$$b^2 = (B_x - A_x)^2 + (B_y - A_y)^2 \tag{2.6}$$

$$c^2 = (B_x - d)^2 + B_y^{\ 2}$$ (2.7)

which provide a pair of simultaneous equations in B_x and B_y.

Subtracting equation 2.6 from 2.7 gives an expression for B_x.

$$B_x = \frac{a^2 - b^2 + c^2 - d^2}{2(A_x - d)} - \frac{2A_y B_y}{2(A_x - d)} = S - \frac{2A_y B_Y}{2(A_x - d)}$$ (2.8)

Substituting equation 2.8 into 2.7 gives a quadratic equation in B_y, which has two solutions corresponding to those in Figure 2.19.

$$B_y^{\ 2} + \left(S - \frac{A_y B_y}{A_x - d} - d \right) - c^2 = 0$$ (2.9)

This can be solved with the familiar expression for the roots of a quadratic equation,

$$B_y = \frac{-Q \pm \sqrt{Q^2 - 4PR}}{2P},$$ (2.10)

where:

$$P = \frac{A_Y^{\ 2}}{(A_x - d)^2} + 1$$ (2.11)

$$Q = \frac{2A_Y(d - S)}{A_x - d}$$ (2.12)

$$R = (d - S^2) - C^2$$ (2.13)

$$S = \frac{a^2 - b^2 + c^2 - d^2}{2(A_x - d)}.$$ (2.14)

Note that the solutions to this equation set can be real or imaginary. If the latter, it indicates that the links cannot connect at the given input angle or at all. Once the two values of B_y are found (if real), they can be substituted into equation 2.8 to find their corresponding x components. The link angles for this position can then be found from

$$\theta_3 = \tan^- \left(\frac{B_Y - A_Y}{B_X - A_X} \right)$$ (2.15)

$$\theta_4 = \tan^- \left(\frac{B_y}{B_x - d} \right).$$ (2.16)

A two-argument arctangent must be used to solve the **above equations** since the angles can be in any quadrant. The above can be encoded in any computer language equation solver, and the value of θ_2 varies over the linkage's usable range to find all corresponding values of the other two link angles.

2.5.3   Complex Algebra Method for Position Analysis

An alternate approach to linkage position analysis creates a vector loop (or loops) around the linkage. The links are represented as position vectors. Figure 2.22 show a slider-crank mechanism with an offset, where the links are drawn as position vectors that form a vector loop. The term offset means that the slider axis extended does not pass through the crank pivot. This is the general case. This linkage could be represented by only three positions vectors, R_2, R_3, and R_s, but one of them (R_s) will be a vector of varying magnitude and angle. It will be easier to use four vectors; R_1, R_2, R_3, and R_4 with R_1 arranged parallel to the axis of sliding and R_4 perpendicular. In effect, the pair of vectors R_1 and R_4 is an orthogonal component of the position vector R_s from the origin to the slider. This loop closes on itself making the sum of the vectors around the loop zero. The lengths of the vectors are the link lengths, which are known.

It simplifies the analysis to arrange one coordinate axis parallel to the axis of sliding. The variable-length, constant-direction vector R_1 then represents the slider position with magnitude d. The vector R_4 is orthogonal to R_1 and defines the constant magnitude offset of the linkage. Note that for the special-case, non-offset version, the vector R_4 will be zero and R_1=R_s. The vectors R_2 and R_3 complete the vector loop. The coupler's position vector R_3 is placed with its root at the slider which then defines its angle θ_3 at point B. This particular arrangement of the position vector leads to a vector loop equation similar to the pin-jointed four-bar example:

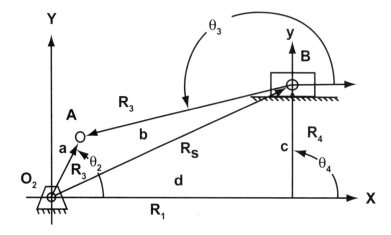

FIGURE 2.22   **Position vector loop for a four-bar slider-crank linkage.**

$R_2 - R_3 - R_4 - R_1 = 0$. Vector R_3 in each case.

The angle $\grave{e}_3$ must always be measured at the root of vector R_3, and in this example it will be convenient to have angle $\grave{e}_4$ at the joint labeled B. Once these arbitrary choices are made it is crucial that the resulting algebraic signs be carefully observed in the equations, or the results will be completely erroneous. Letting the vector magnitudes (link lengths) be represented by a, b, c, and d as shown, we can substitute the complex number equivalents for the position vectors.

$$ae^{j\theta_2} - be^{j\theta_3} - ce^{j\theta_4} - de^{j\theta_1} = 0 \qquad (2.17)$$

Substitute the Euler equations:

$$a(\cos\theta_2 - j\sin\theta_2) - b(\cos\theta_3 + j\sin\theta_3) - c(\cos\theta_4 + j\sin\theta4) \\ - d(\cos\theta1 + j\sin\theta1) = 0. \qquad (2.18)$$

Separate the real and imaginary components.
Real part (x component):

$$a\cos\theta_2 - b\cos\theta_3 - c\cos\theta_4 - d\cos\theta_1 = 0. \qquad (2.19)$$

But:

$$a\cos\theta_2 - b\cos\theta_3 - c\cos\theta_4 - d = 0. \qquad (2.20)$$

Imaginary part (y component):

$$ja\sin\theta_2 - jb\sin\theta_3 - jc\sin\theta_3 - jd\theta_1 = 0. \qquad (2.21)$$

But:
$\theta_1 = 0$, and the j's divide out, so:

$$a\sin\theta_2 - b\sin\theta_3 - c\sin\theta_4 = 0. \qquad (2.22)$$

We want to solve equation 2.18 simultaneously for the two unknowns, link length d and link angle θ_3. The independent variable is crank angle θ_2. Link lengths a and b, the offset c, and angle θ_4 are known. But note that since we set up the coordinate system to be parallel and perpendicular to the axis of the slider block, the angle θ_1 is zero and θ_4 is 90°. Equation 2.22 can be solved for θ_3 and the result substituted into equation 2.19 to solve for d. the solution is :

$$\theta_{3_1} = \arcsin\left(\frac{a\sin\theta_2 - c}{b}\right) \qquad (2.23)$$

$$d = a\cos\theta_2 - b\cos\theta_3. \qquad (2.24)$$

Note that there are again two valid solutions corresponding to the two circuits of the linkage. The arcsine function is multivalued. Its evaluation will give a value between $+90^0$ representing only one circuit of the linkage. The value of d is dependent on the calculated value of θ_3. The value of θ_3 for the second circuit of the linkage can be found from:

$$\theta_{3_2} = \arcsin\left(\frac{-(a\sin\theta_2 - c)}{b}\right) + \pi \cdot \tag{2.25}$$

2.6   A PRACTICAL GUIDE TO USE VARIOUS MECHANISMS

2.6.1   Most Commonly Used Mechanisms in Projects

Wheels

There are three basic wheel types that are used in most wheeled platforms. These are discussed below.

A **fixed standard wheel** has no vertical axis of rotation for steering. Its angle to the chassis is thus fixed, and it is limited to motion back and forth along the wheel plane and rotation around its point of contact with the ground plane. The rolling constraint for this wheel enforces that all motion along the direction of the wheel plane must be accompanied by an appropriate amount of wheel spin so that there is pure rolling at the point of contact. The sliding constraint for this wheel enforces that the component of the wheel's motion orthogonal to the wheel plane must be zero. Fixed standard wheels are used in most of the mobile robot applications.

A **steered standard wheel** differs from the fixed standard wheel only in that there is an additional degree of freedom: the wheel may rotate along a vertical axis passing through the center of the wheel and the ground contact point. This type of wheel is used in most automobiles to enable them to turn. This is

FIGURE 2.23

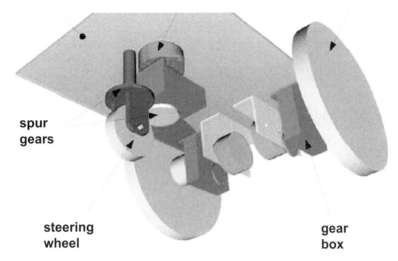

spur
gears

steering
wheel

gear
box

FIGURE 2.24   **A steered wheel.**

discussed in detail in Chapter 4. The disadvantage of this type of wheel is that
the turning radius of the chassis becomes pretty large where this type of wheel
is used. However, it is used mostly in outdoor applications, where the terrain is
somewhat rugged and a very small turning radius is usually not required.

Castor wheels are able to steer around a vertical axis. However, unlike the
steered standard wheel, the vertical axis of rotation in a castor wheel does not

(a)

(b)

FIGURE 2.25

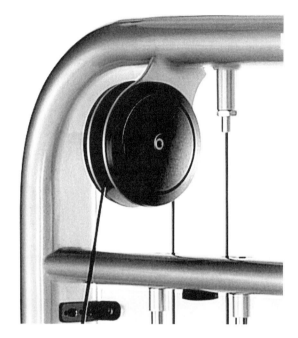

FIGURE 2.26   **Pulleys.**

pass through the ground contact point. The actual kinematics of castor wheels is somewhat complex, however, they do not impose any real constraint on the kinematics of the robot chassis. Castor wheels are usually used to provide support to the structure.

Pulleys can be used to simply change the direction of an applied force or to provide a force/distance tradeoff in addition to a directional change, as shown in Figure 2.26. Pulleys are very flexible because they use ropes to transfer force rather than a rigid object such as a board or a rod. Ropes can be routed through virtually any path. They are able to abruptly change directions in three dimensions without consequence. Ropes can be wrapped around a motor's shaft and either wound up or let out as the motor turns.

Ropes also have the advantage that their performance is not affected by length. If a lever arm were extremely long, then it would be unable to handle the magnitude of forces that a shorter version could withstand. In a lever, to move a given distance next to the fulcrum, the end of the lever must move a distance proportional to its length. As the length of the lever increases, it becomes more likely that the lever will break somewhere along its length.

Figure 2.27 illustrates how a compound pulley 'trades' force for distance through an action/reaction force pair. In a double pulley, as the rope passes over the pulley the force is transmitted entirely but the direction has changed. The

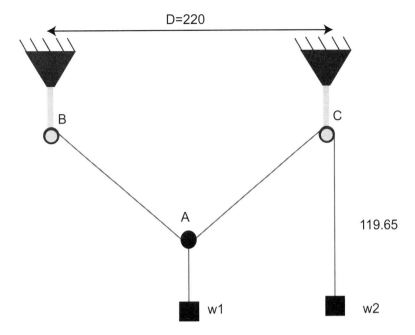

FIGURE 2.27 **How compound pulleys work.**

effort is now pulling up on the left side of the bottom pulley. Now, for a moment forget that the end of the rope is tied to the bottom of the top pulley. The mechanics are the same if the rope is fixed to the ceiling. The important thing is that the end of the rope is immobile. The effort is once again transmitted entirely as the rope passes over the bottom pulley and there is a direction change. The end of the rope is attached to the ceiling so the rope is pulling down on the ceiling with the force of the effort (and half of the force of the load). We assume that the ceiling holds up, so this must mean that there is a force balancing out this downward force. The ceiling pulls up on the rope as a reaction force. This upward force is equal to the effort and now there is an upward force on the right side of the bottom pulley. From the perspective of a free-body diagram the compound pulley system could be replaced by tying two ropes to the load and pulling up on each with a force equal to the effort.

The disadvantages of pulleys, in contrast to machines that use rigid objects to transfer force, are slipping and stretching. A rope will permanently stretch under tension, which may affect the future performance of a device. If a line becomes slack, then the operation of a machine may change entirely. Also, ropes will slip and stick along pulley wheels just like belts. One solution to the problems associated with rope is to use chain. Chain is pliable like rope, and is able to transfer force through many direction changes, but the chain links are inflexible in tension, so that the chain will not stretch. Chains may also be made to fit on gears so that slipping is not a problem.

The Screw

The screw is basically an inclined plane (see Figure 2.28) wrapped around a cylinder. In an inclined plane, a linear force in the horizontal plane is converted to a vertical "lifting" force. With a screw, a rotary force in the horizontal plane is converted to a vertical "lifting" force.

When a wood screw is turned, the threads of the screw push up on the wood. A reaction force from the wood pushes back down on the screw threads and in this way the screw moves down even though the force of turning the screw is in the horizontal plane. Screws are known for high friction, which is why they are used to hold things together. This is true for the LEGO worm gears used in ELEC 201. The friction between these gears and others can take away from the force transmitted through them.

Springs

A favorite device for storing potential energy is the spring. Everything from clocks to catapults makes use of springs. There are two distinctive forms of

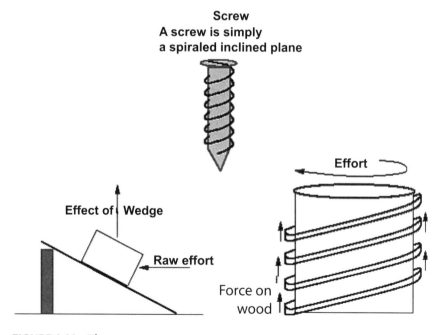

FIGURE 2.28   **The screw.**

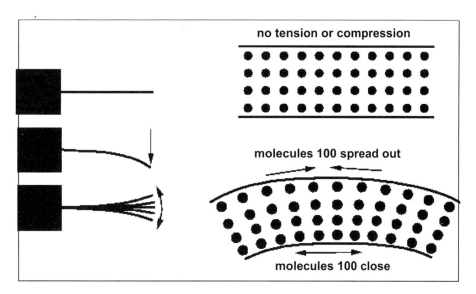

no tension or compression

molecules 100 spread out

molecules 100 close

FIGURE 2.29   **Bar spring.**

springs: the familiar coil and the bending bar. A common use for springs is to return something to its original position. A more interesting application is to use them to measure force—springs in scales. The third use is to store energy. All springs perform all three functions all of the time, but specific devices are built to exploit certain functions of the spring.

A coil spring works for more or less the same reason as a bar spring, it's just in a different shape. To understand a spring, one must zoom in to the microscopic level where molecules interact. Molecules are held together in rigid bodies because of electromagnetic forces. Some of these forces are repulsive, and some of them are attractive. Normally they balance out so that the molecules are evenly spaced within an object; however, by bending a bar, some molecules are forced farther apart and others are shoved closer together. Where the molecules have been spread out, the attractive forces strive to return the original spacing. Where molecules have been forced together, the repulsive forces work to return the object to the original shape.

Rubber Bands

A rubber band is just a kind of spring. A rubber band is slightly more versatile than a metal spring because of its flexibility, just as pulleys are more versatile than their rigid cousin the lever. Using springs in ELEC 201 might take a small amount of imagination, but rubber bands almost scream to be used. There might be several small tasks that a robot performs only once during a round. It would

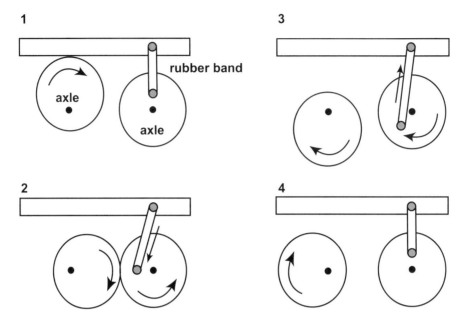

FIGURE 2.30   Using a cam and a rubber band.

not make sense to devote an entire motor to such a task. It's not worth carrying around the extra weight if the task could be accomplished just as well with rubber bands.

Rubber bands also prove useful in the case of repetitive motions. Rather than turning a motor forward then backward then forward and so on, one could make use of a cam and a rubber band to allow the motor to always turn in one direction. Look at the assembly in Figure 2.30 for an example.

Counterweights

Counterweighing is a necessary evil in constructing even a simple robot. Examples of common counterweights are shown in Figure 2.31. If a robot that has been traveling along at a high speed suddenly comes to a halt, there is danger of the robot overturning if the location of the robot's center of mass has not been well placed. The ELEC 201 robots carry around a fairly massive battery, and its placement within the robot's structure is important. When an arm extends, the robot should remain stable. This is accomplished through the use of counterweights.

Counterweighing might also prove useful to raise a bin carrying blocks. Rather than committing an entire motor to raising a bin, a set of counterweights

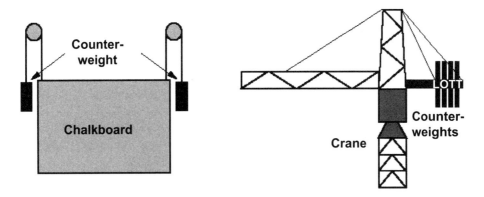

FIGURE 2.31   **Some common counterweights.**

known to be heavier than the bin plus its contents could be suspended until the time when the bin should rise. Of course if a motor was used to take care of the counterweights then no motors have been saved. A motor could be used for more than one task if a mechanical transmission was employed. Another solution would be to use the high-current LED outputs to operate a solenoid.

2.6.2   Use of Different Kinds of Gears and Their Advantages

We will discuss here the various conditions in which different gears are used.

Spur gears are the most widely used gears. They are used for speed reduction and increase. The limitation of these types of gears is that they can transmit motion and power only in an axis parallel to the original direction. They can transmit motion in both directions. That means any gear can be the driving gear.

Helical gears are used in situations similar to spur gears, but are used where more accurate motion transmission is required. The clearance in helical gears is very small as compared to the spur gear of the same dimensions. So these are used for high-speed motion transmission and lesser noise applications.

When motion is to be transmitted in a direction other than that of a parallel to the original axis of the driving gear, bevel gears and worm can be used. Bevel gears can transmit motion in any direction to the original gear direction. But for most practical purposes bevel gears transmit motion at 90 degrees to the original direction. One widely used bevel gear is in an automobile differential system. Bevel gears are used in situations where motion is to be transmitted in both directions. That is, the driving gear can also become the driven gear.

Worm and wheel gears also transmit motion along a perpendicular direction to the original motion. But here the motion transmission is one-way. That is, the

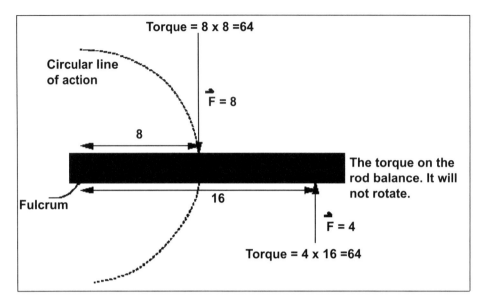

FIGURE 2.32   **Illustrating torque.**

motion can only be transmitted from the worm to the wheel not the other way round. Hence, it is used in situations that require locking in one direction. Many applications in robotics require one-way locking. For example, legged robots and robotic arms. Moreover, the worm and wheel configuration provides much higher gear reduction as compared to the other types of gears for similar dimensions. So, due to these two features of worm and wheel gears, they are the most widely used in robotic applications.

2.6.3   Measuring the Torque of a Motor

Torque

To understand the importance of using a line of action when considering a force, think of a yard stick which has been pinned at the center. The yardstick is free to pivot around its center, so a downward force applied at different places (and thus through different lines of action) will yield different results. Pressing down directly over the pivot does not cause the stick to move or rotate, while pressing down at one end causes the stick to rotate about the pivot. By pressing down at the end, we have applied a torque to the stick and have caused it to rotate.

A torque is a force applied at a distance from a pivot. When describing torques, one must include magnitude, direction, and perpendicular distance from the pivot. For torques the line of action is a circle centered on the pivot. As torque is a product of force and distance, one may be "traded" for the other. By

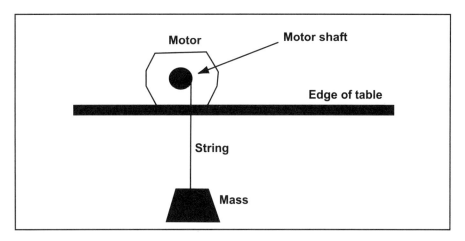

FIGURE 2.33   **Experiment to measure motor torque.**

applying more force closer to the pivot, one may produce the same torque. This concept of "trading" distance traveled/applied for force experienced/applied is key to many simple machines.

A simple experiment can be performed to determine the torque rating of a motor. All that is needed is a motor to be measured, a power supply for the motor, a piece of thread, a mass of known weight, a table, and a ruler. The mass is attached to one end of the thread. The other end of the thread is attached to the motor shaft so that when the motor turns the thread will be wound around the motor shaft. The motor shaft must be long enough to wind the thread like a bobbin.

The motor is put near the edge of a table with the mass hanging over the edge, as illustrated in Figure 2.33. When the motor is powered it will begin winding up the thread and lifting the mass. At first this will be an easy task because the arm movement required to lift the mass is small—the radius of the motor shaft. But soon, the thread will wind around the shaft, increasing the radius at which the force is applied to lift the mass. Eventually, the motor will stall. At this point, the radius of the thread bobbin should be measured. The torque rating of the motor is this radius times the amount of mass that caused the stall.

Alternatively, a LEGO gear and long beam can be mounted on the motor shaft, and a small scale (such as a postage scale) calibrated in grams can be used to measure the force produced by the stalled motor at the end of the lever resting on the scale. The torque in mN-m is given by (force in grams) x (lever length in cm) x (0.09807). The stall current can be measured at the same time. The measurement must be made quickly (1 second) because the large current will heat the motor winding, increasing its resistance, and significantly lowering the current and torque.

Chapter 3 | BASIC ELECTRONICS

In This Chapter

· Introduction to Electronics
· Some Basic Elements
· Steps to Design and Create a Project
· Sensor Design
· Using the Parallel Port of the Computer
· Serial Communication: RS-232
· Using the Microcontroller
· Actuators

3.1 INTRODUCTION TO ELECTRONICS

The concept of electronics is used about electronic components, integrated circuits, and electronic systems. Thirty years ago, no one ever thought of the expansive growth of the electronics, information, and communication technology we have seen the last few decades. Our new digital life is built on the development of miniaturized electronic circuits (microchips) and broadband telephone and data transmission through optical fiber and wireless networks.

The computer has been a common tool both at work and at home. By continued miniaturization of digital electronic components and circuits, the PC and other advanced electronics have been commercially available for people in general. The capacity of computers almost doubles every year. This

FIGURE 3.1   **Introduction.**

expansion is achievable because of tighter packaging of the components onto the microchip.

Modern cars have been exposed by a tremendous development where the main parts of the functions have been controlled by the electronics. The cars are equipped with electronics like airbag systems, ABS brakes, antispinning system, and burglar alarm. Within transportation, we have obtained advanced electronic navigation systems, instrument landing systems for airplanes, and anticollision systems for ships and cars. Automatic toll rings around the largest cities provide money for new roads and attempts for environmentally friendly traffic.

Furthermore, modern electronics have revolutionized medical diagnosis by introducing new techniques like CT (Computer Tomography), MR (Magnetic Resonance), and ultrasonic imaging systems.

Common for the realization of these new technical developments, besides digital circuits, are the sensors that can "feel" sound, light, pressure, temperature, acceleration, etc., and actuators that can "act," i.e,. carry out specific operations like switch on a knob, or transmit sound or light signals.

The base of all the big circuits is capacitors, resistors, and transistors. Any electronic device or equipment generally has these components as its starting point. So before going for the circuits, one needs to has some knowledge about these basic components. Some of these components are discussed below in detail.

3.2   SOME BASIC ELEMENTS

Resistors, capacitors, and transistors are a few basic components that can be seen in most electronic circuits. Resistors come in different values. Their resistance

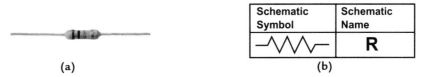

Schematic Symbol	Schematic Name
—⋀⋀⋀—	R

    (a)   (b)

FIGURE 3.2   **(a) A resistor. (b) Schematic symbol & name.**

can be determined by considering the color codes on it. Different components are represented by different symbols. These components having symbolic names like a resistor, is represented by R, whereas a capacitor is represented by C.

3.2.1   Resistors

Fixed Resistors

Resistors are one of the most commonly used components in electronics. As its name implies, resistors resist the flow of electrons. They are used to add resistance to a circuit. The color bands around the resistors are color codes that tell you its resistance value. Recall that a unit of resistance is an ohm.

One last important note about resistors is their wattage rating. You should not use a 1/4-watt resistor in a circuit that has more than 1/4 watt of power flowing.

For example, it is not okay to use a 1/4-watt resistor in a 1/2-watt circuit. However, it is okay to use a 1/2-watt resistor in a 1/4-watt circuit.

The tolerance band indicates the accuracy of the values. A 5% tolerance (gold band) for example, indicates that the resistor will be within 5% of its value. For most applications, a resistor within 5% tolerance should be sufficient.

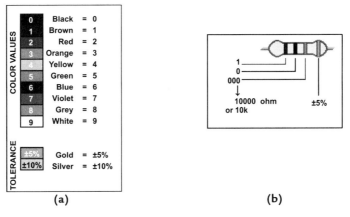

            (a)   (b)

FIGURE 3.3   **Shows the way to calculate the value of a resistor.**

FIGURE 3.4   **A variable resistor.**

To get the value of a resistor, hold the resistor so that the tolerance band is on the right.

The first two color bands from the left are the significant figures—simply write down the numbers represented by the colors. The third band is the multiplier—it tells you how many zeros to put after the significant figures. Put them all together and you have the value.

NOTE   There are resistors with more bands and other types for specific applications. However, 4-band resistors (the ones discussed here) are the most common and should work for most projects.

Variable Resistors

Variable Resistors, or **potentiometers**, often have three terminals and can change resistance easily.

Horizontally Adjustable Presets

These are miniature versions of the standard variable resistor. They are designed to be mounted directly onto the circuit board and adjusted only when the circuit is built. For example, to set the frequency of an alarm tone or the sensitivity of a light-sensitive circuit, a small screwdriver or similar tool is used to adjust the presets.

Preset Symbol

Presets are much cheaper than standard variable resistors so they are sometimes used in projects where a standard variable resistor would normally be used.

Multiturn presets are used where very precise adjustments must be made. The screw must be turned many times (10+) to move the slider from one end of the track to the other, giving very fine control.

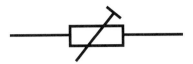

FIGURE 3.5   Symbol for preset.

A variable resistor is a potentiometer with only two connecting wires instead of three. However, although the actual component is the same, it does a very different job. The pot allows us to control the potential passed through a circuit. The variable resistance lets us adjust the resistance between two points in a circuit.

A variable resistance is useful when we don't know in advance what resistor value will be required in a circuit. By using pots as an adjustable resistor we can set the right value once the circuit is working. Controls like this are often called 'presets' because they are set by the manufacturer before the circuit is sent to the customer. They're usually hidden away inside the case of the equipment, away from the fingers of the users!

3.2.2   Capacitors

Capacitors are the second most commonly used component in electronics. They can be thought of as tiny rechargeable batteries—capacitors can be charged and discharged. The amount of charge that a capacitor can hold is measured in Farads or the letter F. However, 1F is too large for capacitors, so microfarads (μF) and picofarads (pF) are used:

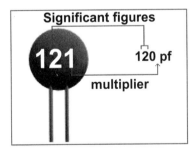

FIGURE 3.6   A capacitor.

micro = 1/1,000,000 and pico = 1/1,000,000,000,000
So, 100,000pF = 0.1μF = 0.0000001F.

We will only be discussing two types of the most commonly used capacitors: ceramic and electrolytic.

Schematic Symbol	Schematic Name
—⟩⊦—	**C**

(a)   (b)

FIGURE 3.7   (a) A ceramic capacitor and (b) its schematic symbol and name.

- A **ceramic capacitor** is brown and has a disc shape. These capacitors are nonpolarized, meaning that you can connect them in any way. To find the value, you simply decode the 3-digit number on the surface of the capacitor. The coding is just like the resistor color codes except that they used numbers instead of colors. The first 2 digits are the significant figures and the third digit is the multiplier. These capacitors are measured in pF.
- **Electrolytic capacitors** have a cylinder shape. These capacitors are polarized so you must connect the negative side in the right place. The value of the resistor as well as the negative side is clearly printed on the capacitor. These capacitors are measured in µF.

3.2.3   Breadboard

Circuits can be modeled to make sure they work the way you want them to. Circuit modeling can be done either using a computer modeling application, or on a prototype board—also called a **breadboard** or **Vero board**—which is a board covered with small sockets into which components can be plugged and connected up.

Figure 3.9 shows a breadboard with holes connected in two long rows at the top and bottom, and columns of five linked holes elsewhere. Electronic components and wires can be simply plugged into the board in order to make any required circuit connections. The top and bottom rows act as power supply channels for the circuit.

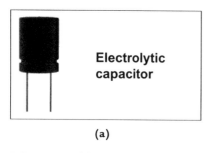

Schematic Symbol	Schematic Name
—⟩⊦+—	**C**

(a)   (b)

FIGURE 3.8   (a) An electrolytic capacitor and (b) its schematic symbol and name.

A breadboard prototype for a 555 mono stable timer circuit might look like this:

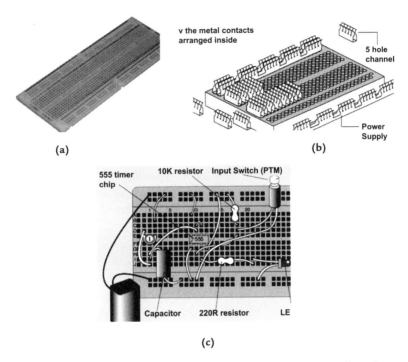

(a)   (b)

(c)

FIGURE 3.9   (a) A breadboard, (b) internal connection of a breadboard, and (c) a symbolic representation of a 555 mono stable timer circuit on a breadboard.

Evaluation and Testing

Evaluation and testing is about making sure that the product stays on track with the design **specification**. You should plan to evaluate and test your project at a number of key stages of design and manufacture. These stages are referred to as **critical control points**.

The critical control points for evaluation and testing an electronic product are:

1. **Initial Design Phase:** Check that you have used the correct value components, and that the various systems work together. These checks can be done using a computer-simulation package.
2. **Breadboard Phase:** Use the **breadboard** to check whether the circuit works properly. Test each part of the circuit using a **millimeter** or **logic probe**.

3. **PCB Layout:** Check that the components are in the correct positions and that you have used the optimum track layout. Make sure that the components are located neatly and that joints are well soldered.

4. **Manufacturing and Packaging Phase:** After manufacture, check that the product conforms to its specification. During packaging, check that the product fits securely in the package, and that any conducting parts are insulated.

5. Finally, the **Analysis Phase:** Look back over the design and making process. Analyze how well it went, noting any modifications and improvements you would make if you were to do it again. These notes are an important part of your design portfolio.

3.2.4   Potentiometer

A potentiometer (or "pot," for short) is a manually adjustable, variable resistor. It is commonly used for volume and tone controls in stereo equipment. On the RoboBoard a 10k pot is used as a contrast dial for the LCD screen, and the RoboKnob of the board is also a potentiometer.

In robotics, a potentiometer can be used as a position sensor. A rotary potentiometer (the most common type) can be used to measure the rotation of a shaft. Gears can be used to connect the rotation of the shaft being measured to the potentiometer shaft. It is easiest to use if the shaft being measured does not need to rotate continuously (like the second hand on a clock), but rather rotates back and forth (like the pendulum on a grandfather clock). Most potentiometers rotate only about 270 degrees; some can be rotated continuously, but the values are the same on each rotation. By using a gear ratio other than 1:1, the position of a shaft that rotates more than 270 degrees can be measured.

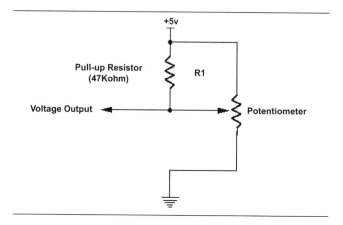

FIGURE 3.10   **Potentiometer circuit.**

FIGURE 3.11   **Diodes.**

A potentiometer connected to a shaft and a lever can also be used to determine the distance to a wall and to make the robot follow a path parallel to the wall. The lever, perhaps with a small wheel on the end, would extend from the side of the robot and contact the wall; a rubber band would provide a restoring force. If the robot moved closer to the wall, the lever would pivot, turning the shaft and the potentiometer. The control program would read the resulting voltage and adjust the robot steering to keep the voltage constant.

Electrical Data

Potentiometers have three terminals. The outer two terminals are connected to a resistor and the resistance between them is constant (the value of the potentiometer). The center terminal is connected to a contact that slides along the resistance element as the shaft is turned, so the resistance between it and *either* of the other terminals varies (one increases while the other decreases).

The assembly instructions suggest wiring the potentiometer in the voltage divider configuration, with the on-board pull-up resistor in parallel with one of the potentiometer's two effective resistances. This will yield readings of greater precision (although they will not be linear) than if the pot were used as a two-terminal variable resistor. You may want to try different circuits to determine which works best for your application.

3.2.5 Diodes

In electronics, a diode is a component that restricts the direction of movement of charge carriers. Essentially, it allows an electric current to flow in one direction, but blocks it in the opposite direction. Today the most common diodes are made from semiconductor materials such as silicon or germanium.

Semiconductor Diodes

Most modern diodes are based on semiconductor P-N junctions. In a P-N diode, conventional current can flow from the P-type side (the anode) to the N-type

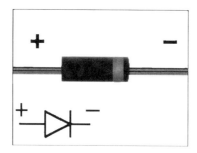

FIGURE 3.12   **Diode schematic symbol.**

side (the cathode), but not in the opposite direction. A semiconductor diode's current-voltage, or I-V, characteristic curve is ascribed to the behavior of the so-called depletion layer or depletion zone which exists at the P-N junction between the differing semiconductors. When a P-N junction is first created, conduction band (mobile) electrons from the N-doped region diffuse into the P-doped region where there is a large population of holes (places for electrons in which no electron is present) with which the electrons "recombine." When a mobile electron recombines with a hole, the hole vanishes and the electron is no longer mobile. Thus, two charge carriers have vanished. The region around the P-N junction becomes depleted of charge carriers and thus behaves as an insulator.

However, the depletion width cannot grow without limit. For each electron-hole pair that recombines, a positively charged dopant ion is left behind in the N-doped region, and a negatively charged dopant ion is left behind in the P-doped region. As recombination proceeds and more ions are created, an increasing electric field develops through the depletion zone which acts to slow and then finally stop recombination. At this point, there is a 'built-in' potential across the depletion zone.

If an external voltage is placed across the diode with the same polarity as the built-in potential, the depletion zone continues to act as an insulator preventing a significant electric current. This is the reverse bias phenomenon. However, if the polarity of the external voltage opposes the built-in potential, recombination can once again proceed resulting in substantial electric current through the P-N junction. For silicon diodes, the built-in potential is approximately 0.6 V. Thus, if an external current is passed through the diode, about 0.6 V will be developed across the diode such that the P-doped region is positive with respect to the N-doped region and the diode is said to be 'turned on' as it has a forward bias.

A diode's I-V characteristic can be approximated by two regions of operation. Below a certain difference in potential between the two leads, the depletion layer has significant width, and the diode can be thought of as an open

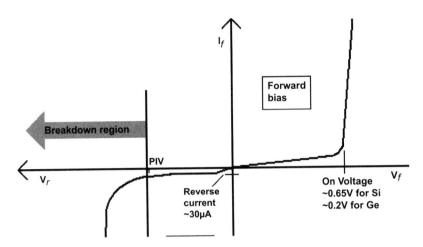

FIGURE 3.13   I-V characteristics of a P-N junction diode (not to scale).

(nonconductive) circuit. As the potential difference is increased, at some stage the diode will become conductive and allow charges to flow, at which point it can be thought of as a connection with zero (or at least very low) resistance. More precisely, the transfer function is logarithmic, but so sharp that it looks like a corner on a zoomed-out graph.

In a normal silicon diode at rated currents, the voltage drop across a conducting diode is approximately 0.6 to 0.7 volts. The value is different for other diode types—Schottky diodes can be as low as 0.2 V and light-emitting diodes (LEDs) can be 1.4 V or more (blue LEDs can be up to 4.0 V).

Referring to the I-V characteristics image, in the reverse bias region for a normal P-N rectifier diode, the current through the device is very low (in the μA range) for all reverse voltages up to a point called the peak-inverse-voltage (PIV). Beyond this point a process called reverse breakdown occurs which causes the device to be damaged along with a large increase in current. For special-purpose diodes like the avalanche or zener diodes, the concept of PIV is not applicable since they have a deliberate breakdown beyond a known reverse current such that the reverse voltage is "clamped" to a known value (called the zener voltage or breakdown voltage). These devices, however, have a maximum limit to the current and power in the zener or avalanche region.

Zener Diode

Refer to the characteristic curve of a typical rectifier (diode) in Figure 3.14. The forward characteristic of the curve we have previously described above in the diode section. It is the reverse characteristics we will discuss here.

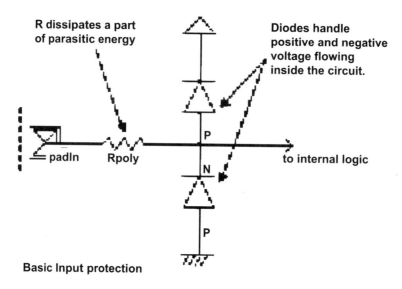

R dissipates a part of parasitic energy

Diodes handle positive and negative voltage flowing inside the circuit.

padIn   Rpoly

P

N

P

to internal logic

Basic Input protection

FIGURE 3.14   Current-voltage characteristics of a typical P-N junction.

Notice that as the reverse voltage is increased the leakage current remains essentially constant until the breakdown voltage is reached where the current increases dramatically. This breakdown voltage is the zener voltage for zener diodes. While for the conventional rectifier or diode it is imperative to operate below this voltage; the zener diode is intended to operate at that voltage, and so finds its greatest application as a voltage regulator.

The basic parameters of a zener diode are:
(a) Obviously, the zener voltage must be specified. The most common range of zener voltage is 3.3 volts to 75 volts; however voltages out of this range are available.
(b) A tolerance of the specified voltage must be stated. While the most popular tolerances are 5% and 10%, more precision tolerances as low as 0.05% are available. A test current (Iz) must be specified with the voltage and tolerance.
(c) The power-handling capability must be specified for the zener diode. Popular power ranges are: 1/4, 1/2, 1, 5, 10, and 50 watts.

Varactor Diode

The varactor diode symbol is shown in Figure 3.15 with a diagram representation.

When a reverse voltage is applied to a P-N junction, the holes in the P-region are attracted to the anode terminal and electrons in the N-region are attracted to the cathode terminal creating a region where there is little current.

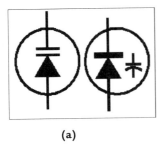
(a)

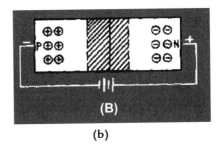
(b)

FIGURE 3.15   (a) Varactor diode symbol and (b) its diagram representation.

This region, the depletion region, is essentially devoid of carriers and behaves as the dielectric of a capacitor.

The depletion region increases as the reverse voltage across it increases; and since capacitance varies inversely as dielectric thickness, the junction capacitance will decrease as the voltage across the P-N junction increases. So by varying the reverse voltage across a P-N junction the junction capacitance can be varied. This is shown in the typical varactor voltage-capacitance curve below in Figure 3.16.

Notice the nonlinear increase in capacitance as the reverse voltage is decreased. This nonlinearity allows the varactor to also be used as a harmonic generator.

Major varactor considerations are:

(a) Capacitance value
(b) Voltage
(c) Variation in capacitance with voltage
(d) Maximum working voltage
(e) Leakage current

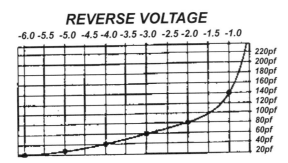

FIGURE 3.16   Varactor voltage-capacitance curve.

Tunnel Diode and Back Diode

Tunnel Diode

A tunnel diode is a semiconductor with a negative resistance region that results in very fast switching speeds, up to 5 GHz. The operation depends upon a quantum mechanic principle known as "tunneling" wherein the intrinsic voltage barrier (0.3 volt for germanium junctions) is reduced due to doping levels which enhance tunneling. Referring to the curves below, superimposing the tunneling characteristic upon a conventional P-N junction, we show in Figure 3.17:

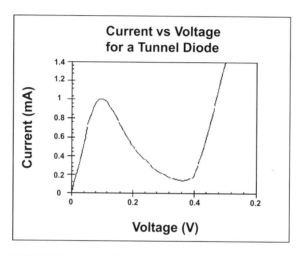

FIGURE 3.17   Combination of tunneling current and conventional P-N junction current resulting in a composite characteristic which is the tunnel diode characteristic curve.

The negative resistance region is the important characteristic for the tunnel diode. In this region, as the voltage is increased, the current decreases; just the opposite of a conventional diode. The most important specifications for the tunnel diode are the Peak Voltage (Vp), Peak Current (Ip), Valley Voltage (Vv), and Valley Current (Iv).

Back Diode

A back diode is a tunnel diode with a suppressed Ip and so approximates a conventional diode characteristic. See the comparison in the figures below:

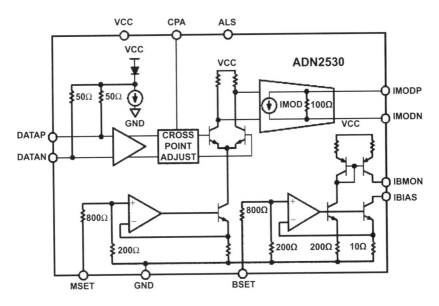

FIGURE 3.18

TABLE 3.1	Typical Tunnel Diodes Supplied by American Microsemiconductor								
Part Number	I_p Peak Point Current (mA)	I_v Valley Point Current Max.	C Capacitance Max. (pF)	V_p Peak Point Voltage Typ. (mV)	V_v Valley Voltage Typ. (mV)	V_{fp} Forward Peak Voltage Typ. (GHz)	R_s Series Resist. Max. (ohms)	-G Negative Conductance (mhosx-10-3)	f_{RO} Resistive Cutoff Frequency Typ.
1N3712	1.0 + 10%	0.18	10	65	350	500	4.0	8 Typ.	2.3
1N3713	1.0 + 2.5%	0.14	5	65	350	510	4.0	8.5 + 1	3.2
1N3714	2.2 + 10%	0.48	25	65	350	500	3.0	18 Typ	2.2
1N3715	2.2 + 2.5%	0.31	10	65	350	510	3.0	19 + 3	3.0
1N3716	4.7 + 10%	1.04	50	65	350	500	2.0	40 Typ.	1.8
1N3717	4.7 + 2.5%	0.60	25	65	350	510	2.0	41 + 5	3.4
1N3718	10.0 + 10%	2.20	90	65	350	500	1.5	80 Typ.	1.6
1N3719	10.0 + 2.5%	1.40	50	65	350	510	1.5	85 + 10	2.8
1N3720	22.0 + 10%	4.80	150	65	350	500	1.0	180 Typ.	1.6
1N3721	22.0 + 2.5%	3.10	100	65	350	510	1.0	190 + 30	2.6

Part Number	I_p Peak point current (mA)	I_v Valley Point Current (mA)	C Capaci- tance Max. (pF) Max. (mV)	V_p Peak Point Voltage (mV)	V_v Valley Voltage Typical I (mV)	V_{fp} Forward Voltage Typica Typical	R_s Series Resist. Typical (ohms)	T Rise Time Typical (psec.)
TABLE 3.2 Typical Ultra-high-speed Switching Tunnel Diodes Supplied by American Microsemiconductor								
TD-261	2.2 ± 10%	0.31	3.0	70	390	500-700	5.0	430
TD-261A	2.2 ± 10%	0.31	1.0	80	390	500-700	7.0	160
TD-262	4.7 ± 10%	0.60	6.0	80	390	500-700	3.5	320
TD-262A	4.7 ± 10%	0.60	1.0	90	400	500-700	1.7	350
TD-263	10.0 ± 10%	1.40	9.0	75	400	500-700	1.7	350
TD-263A	10.0 ± 10%	1.40	5.0	80	410	520-700	2.0	190
TD-263B	10.0 ± 10%	1.40	2.0	90	420	550-700	2.5	68
TD-264	22.0 ± 10%	3.80	18.0	90	425	600 Typ.	1.8	185
TD-264A	22.0 ± 10%	3.80	4.0	100	425	550-700	2.0	64
TD-265	50.0 ± 10%	8.50	25.0	110	425	625 Typ.	1.4	100
TD-265A	50.0 ± 10%	8.50	5.0	130	425	640 Typ.	1.5	35
TD-266	100 ± 10%	17.50	35.0	150	450	650 Typ.	1.1	57
TD-266A	100 ± 10%	17.50	6.0	180	450	650 Typ.	1.2	22

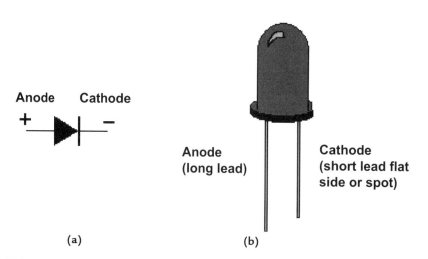

(a)   (b)

FIGURE 3.19   (a) LED symbol. (b) LED diagram.

The reverse breakdown for tunnel diodes is very low, typically 200 mV, and the TD conducts very heavily at the reverse breakdown voltage. Referring to the BD curve, the back diode conducts to a lesser degree in a forward direction. It is the operation between these two points that makes the back diode important. Forward conduction begins at 300 mV (for germanium) and a voltage swing of only 500 mV is required for full-range operation.

3.2.6 LEDs

Definition: Light Emitting Diodes (LEDs) are compound semiconductor devices that convert electricity to light when biased in the forward direction. Because of its small size, ruggedness, fast switching, low power, and compatibility with integrated circuitry, LED was developed for many indicator-type applications.

Today, advanced high-brightness LEDs are the next generation of lighting technology and are currently being installed in a variety of lighting applications. As a result of breakthroughs in material efficiencies and optoelectronic packaging design, LEDs are no longer used in just indicator lamps. They are used as a light source for illumination for monochromatic applications such as traffic signals, brake lights, and commercial signage.

LED Benefits

- Energy efficient
- Compact size
- Low wattage
- Low heat
- Long life
- Extremely robust
- Compatible with integrated circuits

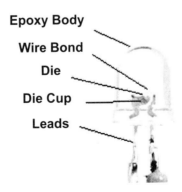

FIGURE 3.20 **Parts of an LED.**

LED Structure

- Chip
- Lead frame
- Gold wire
- Epoxy resin (plastic mold package)
- Cathode
- Anode

TABLE 3.3	Semiconductors for LEDs		
General	**GaP**	**GaN**	**GaAs**	**GaA1As**
	Green, Red	Blue	Red, Infrared	Red, Infrared
Super	**GaAlAs**	**InAsP**	**GaN**	**InGaNGaP**
	Red	Yellow, Red	Blue	Green Green
Ultra	**GaAlAs**	**InGaAlP**	**GaN**	**InGaN**
	Red	Yellow, Orange, Red	Blue	Green

Classification: Classification of LEDs are defined by spectrum.
(i) **Visible LED:** Based on max. spectrum, produces red, orange, yellow, green, blue, and white.
(ii) **Infrared LED:** (IR LED).

Applications of LEDs

Visible LED: General-purpose application in various industries including indication devices for electronic appliances, measuring instruments, etc.

Bi-color (dual color) LED: Charger for cellular phones, showcase boards, traffic boards on highways, etc.

High & Ultra Brightness LED: Full-color display for indoor/outdoor, automotive signal lamps, high-mount lamps, indoor lamps, traffic signal lamps, etc.

Infrared LED: With high output capacity, IR LED is used in remote controls, IrDa (Infrared Data Storage Devices), etc.

3.2.7 Transistors

A semiconductor device consisting of two P-N junctions formed by either a P-type or N-type semiconductor between a pair of opposite types is known as a transistor.

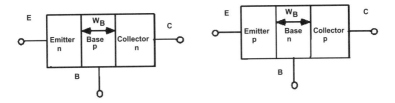

FIGURE 3.21

A transistor in which two blocks of N-type semiconductors are separated by a thin layer of P-type semiconductor is known as an NPN transistor.

A transistor in which two blocks of P-type semiconductors are separated by a thin layer of N-type semiconductor is known as a PNP transistor.

The three portions of a transistor are the emitter, base, and collector, shown as E, B, and C respectively in Figure 3.21.

The section of the transistor that supplies a large number of majority carriers is called the emitter. The emitter is always forward biased with respect to the base so that it can supply a large number of majority carriers to its junction with the base. The biasing of the emitter base junction of an NPN and PNP transistor is shown in Figure 3.22. Since the emitter is to supply or inject a large amount of majority carriers into the base, it is heavily doped but moderate in size.

The section on the other side of the transistor that collects the major portion of the majority carriers supplied by the emitter is called the collector. The collector base junction is always reverse biased. Its main function is to remove majority carriers (or charges) from its junction with the base. The biasing of collector base junctions of an NPN transistor and a PNP transistor is shown in Figure 3.21 above. The collector is moderately doped but larger in size so that it can collect most of the majority carriers supplied by the emitter.

The middle section, which forms two P-N junctions between the emitter and collector, is called the base. The base forms two circuits, one input circuit with emitter and the other an output circuit with collector. The base emitter junction is forward biased providing low resistance for the emitter circuit. The base collector circuit is reversed biased, offering a high-resistance path to the collector circuit. The base is lightly doped and very thin so that it can pass on most of the majority carriers supplied by the emitter to the collector.

Operation of an NPN Transistor

An NPN transistor circuit is shown in Figure 3.22. The emitter base junction is forward biased while the collector base junction is reversed biased. The forward biased voltage v_{eb} is quite small, where as the reversed biased voltage v_{cb} is considerably high.

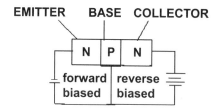

FIGURE 3.22

As the emitter base junction is forward biased, a large number of electrons (majority carriers) in the emitter (N-type region) are pushed toward the base. This constitutes the emitter current i_e. When these electrons enter the P-type material (base), they tend to combine with holes. Since the base is lightly doped and very thin, only a few electrons (less than 5%) combine with holes to constitute base current i_b. The remaining electrons (more than 95%) diffuse across the thin base region and reach the collector space charge layer. These electrons then come under the influence of the positively based N-region and are attracted or collected by the collector. This constitutes collector current i_c.

Thus, it is seen that almost the entire emitter current flows into the collector circuit. However, to be more precise, the emitter current is the sum of the collector current and base current i.e.,

$$i_e = i_c + i_b.$$

Operation of a PNP Transistor

A PNP transistor circuit is shown in Figure 3.23 below. The emitter base junction is forward biased while the collector base junction is reverse biased. The forward-biased voltage v_{eb} is quite small, where as the reverse-biased voltage v_{cb} is considerably high.

As the emitter base junction is forward biased, a large number of holes (majority carriers) in the emitter (P-type semiconductor) are pushed toward the base. This constitutes the emitter current i.e., when these electrons enter the N-type material (base), they tend to combine with electrons. Since the base is lightly

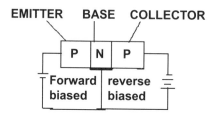

FIGURE 3.23

doped and very thin, only a few holes (less than 5%) combine with electrons to constitute base current $\mathbf{i_b}$. The remaining holes (more than 95%) diffuse across the thin base region and reach the collector space charge layer. These holes then come under the influence of the negatively based P-region and are attracted or collected by the collector. This constitutes collector current $\mathbf{i_c}$.

Thus, it is seen that almost the entire emitter current flows into the collector circuit. However, to be more precise, the emitter current is the sum of the collector current and base current i.e.,

$$i_e = i_c + i_b.$$

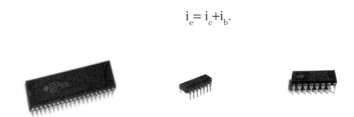

FIGURE: 3.24   Integrated chips

3.2.8   Integrated Circuits

Integrated circuits are miniaturized electronic devices in which a number of active and passive circuit elements are located on or within a continuous body of material to perform the function of a complete circuit. Integrated circuits have a distinctive physical circuit layout, which is first produced in the form of a large-scale drawing and later reduced and reproduced in a solid medium by high-precision electrochemical processes. The term "integrated circuit" is often used interchangeably with such terms as microchip, silicon chip, semiconductor chip, and microelectronic device.

Overview

- ICs, often called "chips," come in several shapes and sizes.
- Most common are 8-, 14-, or 16-pin dual in-line (dil) chips.
- ICs can be soldered directly into printed circuit boards, or may plug into sockets which have already been soldered into the board.
- When soldering, ensure that the IC (or the socket) is the correct way round and that no pins have been bent underneath the body.
- When fitting new ICs it is often necessary to bend the pins in slightly, in order to fit it into the board (or socket).
- Some ICs are damaged by the static electricity that most people carry on their bodies. They should be stored in conductive foam or wrapped in tin foil. When handling them, discharge yourself periodically by touching some metalwork which is earthed, such as a radiator.

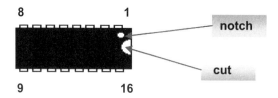

FIGURE 3.25   Parts of a 16-pin chip.

Pin Numbering on a Typical IC

The value of the output voltage from simple power supplies is often not accurate enough for some electronic circuits.

The power supply voltage can also vary due to changes in the main supply, or variations in the current taken by the load.

3.2.9   Some Lab Components

While working with electronic circuits we generally come across so many electronic components that one needs to know. Some of the components that are most common are described below:

IC 7805

The 7805 supplies 5 volts at 1 amp maximum with an input of 7–25 volts.
The 7812 supplies 12 volts at 1 amp with an input of 14.5–30 volts.
The 7815 supplies 15 volts at 1 amp with an input of 17.5–30 volts.
The 7824 supplies 24 volts at 1 amp with an input of 27–38 volts.

The 7905, 7912, 7915, and 7924 are similar but require a negative voltage in and give a negative voltage out.

Note that the electrolytic 10 uF must be reversed for negative supplies. Ensure that the working voltage of this component is sufficient. Say 25 V for the 5-, 12-, and 15-volt supplies and 63 V for the 24-volt supply.

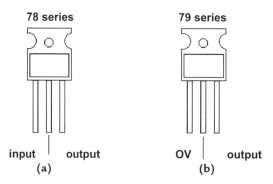

FIGURE 3.26   (a) 78 series. (b) 79 series voltage regulators.

FIGURE 3.27 ULN 2803.

The other two capacitors can be 100 nF/100 volt working. The 78L series can supply 100 mA and the 78S can supply 2 amps.

Eight Darlington Arrays

High-voltage High-current Darlington Transistor Array

- Eight Darlingtons with common emitters.
- Output current to 500 mA.
- Output voltage to 50 V.
- Integral suppression diodes.
- Output can be programmed.

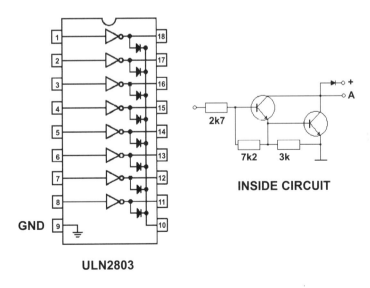

FIGURE 3.28 ULN 2803 (pin connection).

■ Inputs pinned opposite outputs to simplify board layout.
■ Versions for all popular logic families.

Description

The ULN2801A–ULN2805A each contain eight Darlington transistors with common emitters and integral suppression diodes for inductive loads. Each Darlington features a peak load current rating of 600 mA (500 mA continuous) and can withstand at least 50 V in the off state. Outputs may be paralleled for higher current capability.

The output of the ULN2803 is "inverted." This means that a HIGH at the input becomes a LOW at the corresponding output line. E.g., if the motor line connected to pin 1 goes HIGH, pin 18 on the ULN2803 will go LOW (switch off).

The ULN2803 is described as an "**8-line driver**." This means that it contains the circuitry to control eight individual output lines, each acting independently of the others. The IC can be thought of as an 8-line 'black box.'

LM 324 IC

FIGURE 3.29   **14-pin DIP** .

LM324–Quad Operational Amplifier

■ The LM 324 is a **QUAD OP-AMP.**
■ Minimum supply voltage 6 V
■ Maximum supply voltage 15 V
■ Max current per output 15 mA
■ Maximum speed of operation 5 MHz

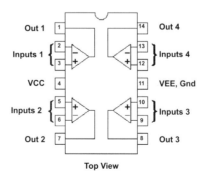

FIGURE 3.30   **Pin diagram of ULN 2803.**

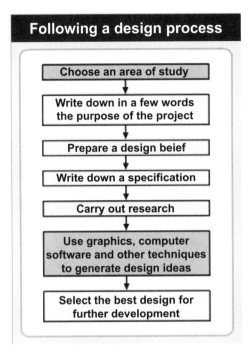

FIGURE 3.31

3.3   STEPS TO DESIGN AND CREATE A PROJECT

A design procedure is a series of steps which guide you through any electronic design-and-make process. Sticking to the procedure will help deliver a first-class product.

Once you have defined the purpose of your project, there are two important documents you need to write. These are:

■ The **design brief** is a short statement of the problem to be solved. The brief should outline the design problem you are tackling, perhaps including one or two of the envisaged design features.

■ The **design specification** is a longer document, including full details of the functional and design features of the finished electronic product as well as information on weight and size, maintenance, cost, and safety.

The specification for an electronic product should include electronic factors such as component details, maximum working voltages, maximum currents, and temperature or frequency ranges.

Ergonomics and Aesthetics

The factors which make a product efficient, safe, and comfortable to use are called **ergonomics**. Considerations of style—the things which make a product look and feel good—are called **aesthetics**. You need to consider both ergonomic and aesthetic factors when planning your designs.

When designing circuits, for example, ensure that switches and other control components are placed so that they can be easily reached, and that output components such as LEDs can be easily seen.

A product's style is a more subjective matter, as different people may have different ideas of what looks good. Think about contemporary style, about what is currently fashionable, when designing your product. You may not want to follow the fashion, but you still need to know what it is!

3.4   SENSOR DESIGN

Without sensors, a robot is just a machine. Robots need sensors to deduce what is happening in their world and to be able to react to changing situations. This section introduces a variety of robotic sensors and explains their electrical use and practical application. The sensor applications presented here are not meant to be exhaustive, but merely to suggest some of the possibilities.

Sensors as Transducers

The basic function of an electronic sensor is to measure some feature of the world, such as light, sound, or pressure and convert that measurement into an electrical signal, usually a voltage or current. Typical sensors respond to stimuli by changing their resistance (photocells), changing their current flow (phototransistors), or changing their voltage output (the Sharp IR sensor). The electrical output of a given sensor can easily be converted into other electrical representations.

Analog and Digital Sensors

There are two basic types of sensors: *analog* and *digital*. The two are quite different in function, in application, and in how they are used with the Robo-Board. An analog sensor produces a *continuously varying* output value over its range of measurement. For example, a particular photocell might have a resistance of 1k ohm in bright light and a resistance of 300k ohm in complete darkness. Any value between these two is possible depending on the particular light level present. Digital sensors, on the other hand, have only two states, often called "on" and "off." Perhaps the simplest example of a digital sensor is the touch switch. A typical touch switch is an open circuit (infinite resistance) when it is not pressed, and a short circuit (zero resistance) when it is depressed.

Some sensors that produce a digital output are more complicated. These sensors produce *pulse trains* of transitions between the 0-volt state and the 5-volt state. With these types of sensors, the frequency characteristics or shape of this pulse train convey the sensor's measurement. An example of this type of sensor is the Sharp modulated infrared light detector. With this sensor, the actual element-measuring light is an analog device, but signal-processing circuitry is integral to the sensor producing a digital output.

Sensor Inputs on the RoboBoard

The RoboBoard contains input ports for both analog and digital sensors. While both types of ports are sensitive to voltage, each type interprets the input voltage differently and provides different data to the microprocessor. The analog ports measure the voltage and convert it to a number between 0 and 255, corresponding to input voltage levels between 0 and 5 volts. The conversion scale is linear, so a voltage of 2.5 volts would generate an output value of 127 or 128. The digital ports, however, convert an input voltage to just two output values, zero and one. If the voltage on a digital port is less than 2.5 volts, the output will be 0, while if the input is greater than 2.5 volts, the output will be 1. Thus, the conversion is very nonlinear.

Reading Sensor Inputs

The C library function analog (port-#) is used to return the value of a particular analog sensor port. For example, the IC statement

$$val = analog(27);$$

sets the value of the variable val equal to the output of port #27.

Many devices used as digital sensors are wired to be *active low*, meaning that they generate 0 volts when they are active (or true). The digital inputs on the RoboBoard have a pull-up resistor that makes the voltage input equal to 5 volts when nothing is connected. A closed or depressed touch switch connected to a digital port would change that voltage to 0 volts by shorting the input to ground. The resulting outputs: open switch = 1, and closed switch = 0, are the logical opposite of what we usually want. That is, we would prefer the output of the digital port to have value 0 or False normally, and change to 1 or True only when the switch hit something (like a wall or another robot) and was depressed. The IC library function digital (port-#), used to read a True-or-False value associated with a particular sensor port, performs this logical inversion of the signal measured on a digital port. Hence, the depressed touch switch (measuring 0 volts on the hardware) causes the digital () function to return a 1 (logic True) or logical True value.

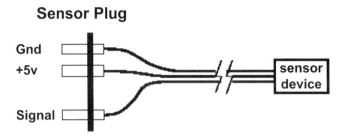

FIGURE 3.32   Generic sensor wiring.

For example, the C statement

$$\text{if (digital(2)) do\_it();}$$

returns a True value (the number 1) and calls the function do_it() if the value at port #2 was 0 volts (indicating a depressed switch).

Connector Plug Standard

The standard plug configuration used to connect sensors to the RoboBoard is shown in Figure 3.32. Notice that the plug is asymmetric (made by removing one pin from a four-pin section of the male header), and is therefore *polarized*. The plug can only be inserted in the RoboBoard port in one orientation, so once the plug is wired correctly, it cannot be inserted into a sensor port backward. This makes the plug much easier to use correctly, but, of course, if you wire it incorrectly, you must rewire it since you cannot turn the plug around.

Generally, the sensor is connected to the plug with three wires. Two of the wires supply 5-volt power from the RoboBoard, labeled "+5v" and "Gnd." The third wire, labeled "Signal" is the voltage output of the sensor. It is the job of the sensor to use the power and ground connections (if necessary) and return its "answer," as a voltage, on the Signal wire.

Sensor Wiring

Figure 3.33 shows a diagram of circuitry associated with each sensor. This circuitry, residing on the RoboBoard, is replicated for each sensor input channel. The key thing to notice is the *pull-up resistor* wired from the sensor input signal leads to the 5-volt power supply.

There are two reasons why this resistor is used. First, it provides a default value for the sensor input—a value when no sensor is plugged in. Many ICs, such as those on the board that read and interpret the sensor voltage, do not perform

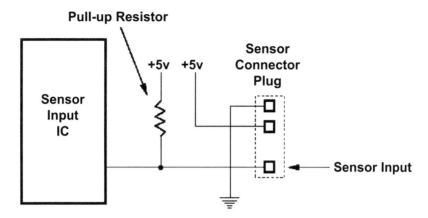

FIGURE 3.33   Sensor input port circuitry.

well when their inputs are left unconnected. With this circuit, when nothing is plugged into the sensor port, the pull-up resistor provides a 5-volt bias voltage on the sensor input line. Thus, the default output value of an analog port is 255, while the default output value of a digital port is 0 or logic False. (Remember that the 5-volt default input value would lead to a digital value of 1 except that this value is inverted by the digital () library function, as explained earlier.)

Second, the pull-up resistor is also used as part of the *voltage divider* circuit required for some of the analog sensors, as explained in the following section. The resistors on the RoboBoard eliminate the need for an additional resistor on each of the sensors.

The Voltage Divider Circuit

Most of the sensors used in the RoboBoard kit make use of the voltage divider circuit shown in Figure 3.34. In the voltage divider, the voltage measured at the common point of the two resistors, V_{out}, is a function of the input voltage, V_{in} (5 volts in this case), and the values of the two resistors, R_1 and R_2. This voltage can be calculated using Ohm's law, $V = I \times R$. The current, I, flowing through the circuit shown in the diagram, is $V_{in} / R_1 + R_2$ (calculated using the rule that series resistances add). Then V_{out}, the voltage drop across R_2, is $R_2 \times i$, which yields the result:

$$V_{out} = V_{in} (R_2 / R_1 + R_2).$$

In ELEC 201 applications, R_1 has a fixed or constant value (as shown in Figure 3.34), while R_2 is the variable resistance produced by the sensor. V_{in} is the positive voltage supply, fixed at 5 volts. Thus, the V_{out} signal can be directly

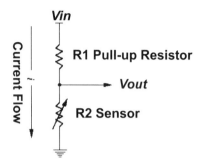

FIGURE 3.34   Voltage divider schematic.

computed from R_2, the resistive sensor. From looking at the equation, it is easy to see that if R_2 is large with respect to R_1, the output voltage will be large, and if R_2 is small with respect to R_1, the output voltage will be small. The minimum and maximum possible voltage values are 0 and 5 volts, matching the RoboBoard input circuitry range.

Tactile Sensors

The primary sensors in the ELEC 201 kit used to detect tactile contact are simple push-button or lever-actuated switches. *Microswitch* is the brand name of a variety of switches that are so widely used that the brand name has become the generic term for this type of switch, which is now manufactured by many companies. A microswitch is housed in a rectangular body and has a very small button (the "switch nub") which is the external switching point. A lever arm on the switch reduces the force needed to actuate the switch (see Figure 3.35). Microswitches are an especially good type of switch to use for making touch sensors.

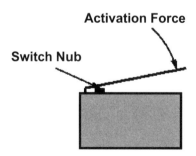

FIGURE 3.35   A typical microswitch.

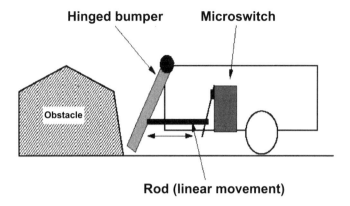

FIGURE 3.36   Robotic platform employing a bumper coupled to a touch sensor.

Often, the switch is simply mounted on a robot so that when the robot runs into something, the switch is depressed, and the microprocessor can detect that the robot has made contact with some object and take appropriate action. However, creative mechanical design of the bumper-switch mechanism is required so that contact with various objects (a wall, a robot, etc.) over a range of angles will be consistently detected. A very sensitive touch bumper can be made by connecting a mechanism as an extension of the microswitch's lever arm, as illustrated in Figure 3.36.

Limit Switch

Touch sensors can also serve as *limit switches* to determine when some movable part of the robot has reached the desired position. For example, if a robot arm is driven by a motor, perhaps using a gear rack, touch switches could detect when the arm reached the limit of travel on the rack in each direction.

Switch Circuitry

Figure 3.37 shows how a switch is wired to a sensor input port. When the switch is open (as it is shown in the diagram), the sensor input is connected to the 5-volt supply by the pull-up resistor. When the switch is closed, the input is connected directly to ground, generating a 0-volt signal (and causing current to flow through the resistor and switch).

Most push-button-style switches are "normally open," meaning that the switch contacts are in the open-circuit position when the switch has not been pressed. Microswitches often have both normally open and normally closed contacts along with a common contact. When wiring a microswitch, it is customary to use the normally open contacts. Also, this configuration is the active-low mode expected by the standard library software used to read the output values

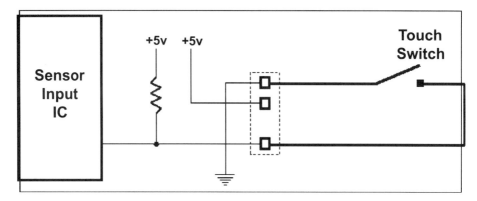

FIGURE 3.37   Touch switch circuit.

from digital sensor ports. However, you can wire the switch differently to perform some special function. In particular, several switches can be wired in series or parallel and connected to a single digital input port. For example, a touch bumper might have two switches, and the robot only needs to know if *either* of them (#1 OR #2) are closed. It takes less code and less time to check just one digital port and to use parallel switch wiring to implement the logic OR function in hardware.

Mercury Tilt Switch

As the name suggests, a mercury tilt switch contains a small amount of mercury inside a glass bulb. The operation of the switch is based on the unique properties of mercury: it is both a conductor and a liquid. When the switch tilts mercury flows to the bottom of the bulb closing the circuit between two metal pins.

The mercury tilt switch can be used in any application to sense inclination. For example, the tilt switch could be used to adjust the position of an arm or ramp. Most thermostats contain a mercury tilt switch mounted on a temperature sensitive spring. Changes in temperature tilt the switch, turning the furnace or air conditioner on or off.

Light Sensors

Measurement of light provides very valuable information about the environment. Often, particular features of the game board (such as goals) are marked by a strong light beacon. The board surface has contrasting lines that can be detected by the difference in the amount of light they reflect. A variety of light sensors are provided in the ELEC 201 kit:

Infrared Reflectance Sensor

This device combines an infrared LED light source and a phototransistor light detector into a single package. The LED and the detector point out of the package, almost parallel to each other. The detector will measure the light scattered or reflected by a surface a short distance away. The package also contains an optical filter (colored plastic) that transmits primarily only the infrared light from the LED; this reduces, but does not eliminate, the sensitivity to ambient light.

Infrared Slotted Optical Switch

This device is similar to the IR reflectance sensor, in that it contains both an infrared source and a filtered infrared phototransistor detector. However, the two are mounted exactly opposite each other with a small, open gap between them. The sensor is designed to detect the presence of an object in the gap that blocks the light.

Modulated Infrared Light Detector

This device senses the presence of infrared light that has been modulated (e.g., blinks on and off) at a particular frequency. These devices are typically used to decode the signals of TV remote controls, but are used in the ELEC 201 application to detect the infrared beacon of the opponent robot.

Photocell

This device is a light-dependent resistor. It is most sensitive to visible light in the red.

Photocells are made from a compound called cadmium sulfide (CdS) that changes resistance when exposed to varying degrees of light. Cadmium sulfide photocells are most sensitive to visible red light, with some sensitivity to other wavelengths.

Photocells have a relatively slow response to changes in light. The characteristic blinking of overhead fluorescent lamps, which turn on and off at the 60 Hertz line frequency, is not detected by photocells. This is in contrast to phototransistors, which have frequency responses easily reaching above 10,000 Hertz and more. Therefore, if both sensors were used to measure the same fluorescent lamp, the photocell would show the light to be always on and the phototransistor would show the light to be blinking on and off.

Photocells are commonly used to detect the incandescent lamp that acts as a contest start indicator. They are also used to find the light beacons marking certain parts of the board, such as the goals. While they can be used to measure the reflectivity of the game board surface if coupled with a light source such as a red LED or an incandescent lamp, the IR reflectance sensors are usually better at this function. Photocells are sensitive to ambient lighting and usually need to be shielded. Certain parts of the game board might be marked with polarized light sources. An array of photocells with polarizing filters at different orientations could be used to detect the polarization angle of polarized light and locate those board features.

The photocell acts as resistor R_2 in the voltage divider configuration. The resistance of a photocell decreases with an increase in illumination (an inverse relationship). Because of the wiring of the voltage divider (the photocell is on the lower side of the voltage divider), an increase in light will correspond to a decrease in sensor voltage and a lower analog value.

Infrared Reflectance Sensor

The infrared reflectance sensor is a small rectangular device that contains a phototransistor (sensitive to infrared light) and an infrared emitter. The amount of light reflected from the emitter into the phototransistor yields a measurement of a surface's reflectance, for example, to determine whether the surface is black or white. The phototransistor has peak sensitivity at the wavelength of the emitter (a near-visible infrared), but is also sensitive to visible light and infrared light emitted by visible light sources. For this reason, the device should be shielded from ambient lighting as much as possible in order to obtain reliable results.

The amount of light reflected from the emitter into the phototransistor yields a measurement of a surface's reflectance (when other factors, such as the distance from the sensor to the surface, are held constant). The reflectance sensor can also be used to measure distance, provided that the surface reflectance is constant. A reflectance sensor can be used to detect features drawn on a surface

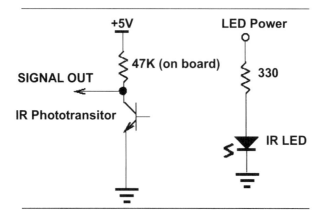

FIGURE 3.38   Phototransistor and infrared emitter circuit.

or segments on a wheel used to encode rotations of a shaft. It is important to remember that the reflectivity measurement indicates the surface's reflectivity at a particular wavelength of light (the near-visible infrared). A surface's properties with respect to visible light may or may not be indicators of infrared light reflectance. In general, though, surfaces that absorb visible light (making them appear dark to the eye) will absorb infrared light as well.

The sensor part (the phototransistor) can be used alone as a light sensor, for example, to detect the starting light, and it is usually much more sensitive than the photocell.

Phototransistor

The light falling on a phototransistor creates charge carriers in the base region of a transistor, effectively providing base current. The intensity of the light determines the effective base drive and thus the conductivity of the transistor. Greater amounts of light cause greater currents to flow through the collector-emitter leads. Because a transistor is an active element having current gain, the phototransistor is more sensitive than a simple photoresistor. However, the increased sensitivity comes at the price of reduced dynamic range. Dynamic range is the difference between the lowest and highest levels that can be measured. The RoboBoard analog sensor inputs have a range of 0–5 volts, and relatively small variations in light can cause the phototransistor output to change through this range. The exact range depends on the circuit used.

As shown in Figure 3.38, the phototransistor is wired in a configuration similar to the voltage divider. The variable current traveling through the resistor causes a voltage drop in the pull-up resistor. This voltage is measured as the output of the device.

Infrared Emitter

The Light Emitting Element (an LED) uses a resistor to limit the current that can flow through the device to the proper value of about 10 milliamps. Normally the emitter is always on, but it could be wired to one of the LED output ports if you wanted to control it separately. In this way you could use the same sensor to detect the starting light (using the phototransistor with the emitter off) and then to follow a line on the board (normal operation with the emitter on).

Infrared Slotted Optical Switch

The infrared slotted optical switch is similar to the infrared reflectance sensor except that the emitter is pointed directly at the phototransistor across a small gap. As the name implies, the slotted optical switch is a *digital* sensor, designed to provide only two output states. The output of the switch changes if something opaque enters the gap and blocks the light path. The slotted optical switch is commonly used to build shaft encoders, which count the revolutions of a shaft. A gear or other type of wheel with holes or slots is placed in the gap between the emitter and detector. The light pulses created by the turning wheel can be detected and counted with special software to yield rotation or distance data. This detector also might be used to detect when an arm or other part of the robot has reached a particular position by attaching a piece of cardboard to the arm so that it entered the gap at the desired arm position.

The slotted optical switch operates in the same fashion as the infrared reflectance sensor, with the exception that a different value of pull-up resistor must be added externally for the particular model of optical switch we use.

Modulated Infrared Light Detector

The modulated infrared light detector is a device that combines an infrared phototransistor with specialized signal-processing circuitry to detect only light that is pulsing at a particular rate. The ELEC 201 kit includes the Sharp GP1U52 sensor, which detects the presence of infrared light modulated (pulsed) at 40,000 Hz. Normal room light, which is not modulated, does not affect the sensor, a big advantage. This type of sensor is used for the remote control function on televisions, VCRs, etc. In ELEC 201 this sensor is used to detect the specially modulated infrared light emitted by the beacon on the opponent robot. The software can distinguish different pulse patterns in order to distinguish between the beacons on the two robots. (In a television remote, different pulse patterns would correspond to different functions, such as changing the channel up or down.)

The principles of operation and use are explained later, when we discuss the circuit used to create the modulated infrared light for the beacon.

Other Sensors

Magnetic Field SensoAvrs

The ELEC 201 kit contains both an analog sensor that provides information about the strength of the magnetic field and a digital sensor, a magnetic switch.

A device called a hall-effect sensor can be used to detect the presence and strength of magnetic fields. The hall-effect sensors have an output voltage even when no magnetic field is present, and the output changes when a magnetic field is present, the direction of change depending on the polarity of the field.

The digital magnetic sensors are simple switches that are open or closed. Internally the switches have an arm made of magnetic material that is attracted to a magnet and moves to short out the switch contacts. These switches are commonly used as door and window position sensors in home security systems. The switch will close when it comes within 1″ of its companion magnet.

TABLE 3.4 Hall-effect sensor test results		
Distance (mm)	Voltage (V)	RoboBoard Reading
0	0.7	35
1	1.55	79
2	1.83	93
3	1.92	98
4	2.03	104
5	2.11	108
6	2.17	111
7	2.21	113
10	2.31	118
15	2.4	123
20	2.47	125
25	2.5	128
30	2.52	129
35	2.54	130
40	2.55	130

Chart shows voltage drop across hall-effect sensor and corresponding reading on the RoboBoard.

Figure 3.39 below provides a visual representation of the hall-effect sensor's range. 2.55 V or 130 on the RoboBoard is the zero reading or when no magneticfield is detected. Fluctuations in voltage may cause a change in the RoboBoard reading by up to two units. For certain detection, look for a change of five units.

For a change of 5 units on the RoboBoard reading, the hall-effect sensor has a range of less than 2.5 cm. This provides a precise tool for determining location designated by a magnet, but offers a limited margin for error.

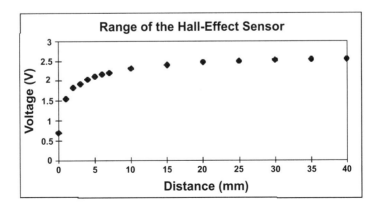

FIGURE 3.39

Either sensor can be used to detect magnets or magnetic strips that may be present on the ELEC 201 game board table. With the magnets typically used on the game board, the hall-effect sensor output voltage changes only a small amount when a field is present. The no-field voltage varies between sensors, but it is very stable for a particular sensor, so the small changes can be detected reliably to determine the presence of a magnet. Hall-effect sensors can be used to make magnetic shaft encoders by mounting a small piece of magnet on a wheel that rotates past the sensor element. Hall-effect sensors can also be used to build a proximity sensor or bounce-free switch, which detects a magnet mounted on a moving component when it is near the sensor element.

Magnetic switches are used in much the same way as a touch switch, except the switch closes when it is near a magnet, instead of when it contacts something. The digital nature of the switch makes it easier to use than the hall-effect sensors, but it may be less sensitive. You should try both.

They can also be used to make an inclination sensor by dangling a magnet above the sensor.

The hall-effect sensor included in the ELEC 201 kit is a digital device that operates from a 5-volt power supply. It uses about 6 mA of current for standard operation. It can sink 250 mA of current into its output, creating logic low. The sensor cannot drive logic high and therefore requires a pull-up resistor for proper operation.

Motor Current Sensor

The motor output drivers of the ELEC 201 RoboBoard contain circuitry that produces an output voltage related to the amount of current being used by a motor. Since the motor current corresponds to the load on the motor, this signal can be used to determine if the motor is stalled and the robot is stuck. The voltage signal depends on a number of factors, including battery voltage, and must be calibrated for each application.

3.5   USING THE PARALLEL PORT OF THE COMPUTER

A port contains a set of signal lines that the CPU uses to send or receive data with other components. Ports are usually used to communicate via modem, printer, keyboard, mouse, etc. In signaling, open signals are "1" and closed signals are "0." A parallel port sends 8 bits and receives 5 bits at a time. The serial port RS-232 sends only 1 bit at a time. However, it is multidirectional. So it can send 1 bit and receive 1 bit at a time. Parallel ports are mainly meant for connecting the printer to the PC. But this port can be programmed for many more applications beyond that.

Parallel ports are easy to program and faster compared to the serial ports. But the main disadvantage is that it needs more number of transmission lines. Because of this reason parallel ports are not used in long-distance communications. You should know the basic difference between working off a parallel port and serial port. In serial ports, there will be two data lines: one transmission and one receiving line. To send data in a serial port, it has to be sent one bit after another with some extra bits like start bit, stop bit, and parity bit to detect errors. But in a parallel port, all the 8 bits of a byte will be sent to the port at a time and an indication will be sent in another line. There will be some data lines, some control, and some handshaking lines in a parallel port.

The D-25 type of female connector is located at the backside of the CPU cabinet and has 25 pins. The pin structure of D-25 is explained in Table 3.5.

TABLE 3.5   Pin Directions and Associated Registers

Pin No (D-Type 25)	SPP Signal	Direction In/out	Register.bit
1*	nStrobe	In/Out	Control.0
2	Data 0	In/Out	Data.0
3	Data 1	In/Out	Data.1
4	Data 2	In/Out	Data.2
5	Data 3	In/Out	Data.3
6	Data 4	In/Out	Data.4
7	Data 5	In/Out	Data.5
8	Data 6	In/Out	Data.6
9	Data 7	In/Out	Data.7
10	nAck	In	Status.7
11*	Busy	In	Status.6
12	Paper-Out / Paper-End	In	Status.5
13	Select	In	Status.4
14*	nAuto-Linefeed	In/Out	Control.1
15	nError / nFault	In	Status.3
16	nInitialize	In/Out	Control.2
17*	nSelect-Printer/ nSelect-In	In/Out	Control.3
18–25	Ground	Gnd	

■ 8 output pins [**D0 to D7**]
■ 5 status pins [**S4 to S7 and S3**]
■ 4 control pins [**C0 to C3**]
■ 8 ground pins [**18 to 25**]

In Figure 3.40, let us see how communication between a PC and printer takes place. The computer places the data in the data pins, and then it makes the strobe

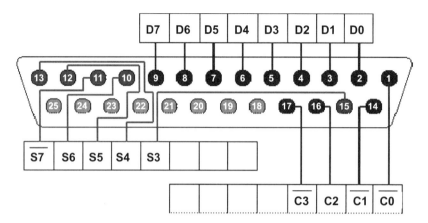

FIGURE 3.40   **Pin configuration of D-25.**

low. When the strobe goes low, the printer understands that there is valid data in the data pins. Other pins are used to send controls to the printer and get the status of the printer; you can understand them by the names assigned to the pins.

To use the printer port for applications other than printing, we need to know how ports are organized. There are three registers associated with an LPT port: data register, control register, and status register. Data register will hold the data of the data pins of the port. That means, if we store a byte of data to the data register, that data will be sent to the data pins of the port. Similarly with control and status registers. Table 3.6 explains how these registers are associated with ports.

TABLE 3.6 Addresses of Data, Control, and Status Registers		
Register	LPT1	LPT2
Data register (Base Address + 0)	0x378	0x278
Status register (Base Address + 1)	0x379	0x279
Control register (Base Address + 2)	0x37a	0x27a

Pins with an * symbol in this table are hardware inverted. That means, if a pin has a 'low' i.e., 0 V, the corresponding bit in the register will have value 1.

Signals with the prefix 'n' are active low. That means that normally these pins will have low value. When it needs to send some indication, it will become high. For example, normally nStrobe will be high, when the data is placed in the port, the computer makes that pin low.

Normally, data, control, and status registers will have the following addresses. We need these addresses in programming later.

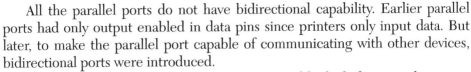

All the parallel ports do not have bidirectional capability. Earlier parallel ports had only output enabled in data pins since printers only input data. But later, to make the parallel port capable of communicating with other devices, bidirectional ports were introduced.

By default, a data port is an output port. To enable the bidirectional property of the port, we need to set the bit 5 of the control register.

To know the details of parallel ports available in your computer, follow this procedure:

- Right click on My Computer, go to "Properties."
- Select the tab Hardware, click Device Manager.
- You will get a tree structure of devices; in that expand "Ports (Com1 & LPT)."
- Double-click on the ECP Printer Port (LPT1) or any other LPT port if available.
- You will get details of the LPT port. Make sure that "Use this Port (enable)" is selected.
- Select tab resourses. In that you will get the address range of the port.

To start programming, you will need a D-25-type male connector.

Programming the Printer Port in DOS

To start programming the port, we will use DOS. In DOS we have commands to access the port directly. But, these programs will not work on the systems based on Windows XP, Windows NT, or higher versions. For security reasons, higher versions of Windows do not allow accessing the port directly. To program the parallel port in these systems, we need to write to the kernel mode driver. In Section 3.5.2, we will discuss programming the parallel port in Windows XP.

When we want to find out whether a particular pin of the port is high or low, we need to input the value of the corresponding register as a byte. In that, we have to find out whether the corresponding bit is high or low using bitwise operators. We can't access the pins individually. So, you need to know basic bitwise operations.

The main bitwise operators that we need are bitwise AND '&' and bitwise OR '|.' To make a particular bit in a byte high without affecting other bits, write a byte with corresponding bit 1 and all other bits 0; OR it with the original byte. Similarly, to make a particular bit low, write a byte with corresponding bit 0 and all other bits 1; AND it with the original byte.

In Turbo C, there are the following functions used for accessing the port:

- outportb(PORTID, data);

- data = inportb(PORTID);
- outport(PORTID, data);
- data = inport(PORTID).

The outport() function sends a word to the port, inport() reads a word from the port. outportb() sends a byte to the port and inportb() reads a byte from the port. If you include the DOS.H header, these functions will be considered as macro, otherwise as functions. Function inport() will return a word having lower byte as the data at PORTID and higher byte as the data at PORTID+2. So, we can use this function to read status and control registers together. inportb() function returns byte at PORTID. outport() writes the lower byte to PORTID and higher byte to PORTID+1. So this can be used to write data and

LISTING 3.1

```
/*    file: ex1.c
            Displays contents of status register of parallel port.
            Tested with TurboC 3.0 and Borland C 3.1 for DOS.
*/
#include"stdio.h"
#include"conio.h"
#include"dos.h"

#define PORT 0x378

void main()
{
        int data;
        clrscr();
        while(!kbhit())
        {
                data=inportb(PORT+1);
                gotoxy(3,10);
                printf("Data available in status register: %3d (decimal),
%3X (hex)\n", data, data);
                printf("\n Pin 15: %d",(data & 0x08)/0x08);
                printf("\n Pin 13: %d",(data & 0x10)/0x10);
                printf("\n Pin 12: %d",(data & 0x20)/0x20);
                printf("\n Pin 11: %d",(data & 0x80)/0x80);
                printf("\n Pin 10: %d",(data & 0x40)/0x40);
                delay(10);
        }
}
```

control together. outportb() function writes the data to PORTID. outport() and outportb() return nothing.

Let us start with inputting first. Here is an example program, copy it and run it in Turbo C or Borland C without anything connected to the parallel port. Then you should see data available in the status register and pin numbers 10, 11, 12, 13, and 15 of the parallel port. Pin 11 (active low) is 0 and all other pins are 1 means it is OK.

To understand bitwise operations: you can find data in pin 15, the value of (data & 0x08) will be 0x08 if bit 3 of the register is high, otherwise:

LISTING 3.2	
bit no. 7654 3210 　　　data : XXXX 1XXX　& 　　　with : 0000 1000　(0x08) 　　　-> 0000 1000　(0x08 -> bit 3 is high)	bit no. 7654 3210 　　　data : XXXX 0XXX　& 　　　with : 0000 1000　(0x08) 　　　-> 0000 0000　(0x00 -> bit 3 is low)

We will use the same logic throughout this section.

Now, take a D-25 male with cables connected to each pin. Short all the pins from 18 to 25, call it as ground. Now you can run the above program and see the change by shorting pins 10, 11, 12, 13, and 15 to ground. We prefer using switches between each input pin and ground. Be careful, do not try to ground the output pins.

To find out the availability of ports in a computer programmatically, we will use the memory location where the address of the port is stored.

TABLE 3.7					
0x408	0x409	0x40a	0x40b	0x40c	0x40d
LPT1 lowbyte	**LPT1** highbyte	**LPT2** lowbyte	**LPT2** highbyte	**LPT3** lowbyte	**LPT3**highbyte

If the following code is run in Turbo C or Borland C, the addresses of available ports can be seen. See Listing 3.3.

Next we will check output pins. To check the output, we will use LEDs. We have driven LEDs directly from the port. But it is preferred to connect a buffer to prevent excessive draw of current from the port. Connect an LED in series with a resister of 1KW or 2.2KW between any of the data pins (2 to 9) and ground. With that, if you run the program in Listing 3.4, you should see the LED blinking with app. 1 sec frequency.

LISTING 3.3

```
/*PortAdd.c
To find availability and addresses of the lpt ports in the computer.
*/
#include <stdio.h>
#include <dos.h>
void main()
{
        unsigned int far *ptraddr;
        /* Pointer to location of Port Addresses */
        unsigned int address;      /* Address of Port */
        int a;
        ptraddr=(unsigned int far *)0x00000408;
        clrscr();

        for (a = 0; a < 3; a++)
        {
                address = *ptraddr;
                if (address == 0)
                        printf("No port found for LPT%d \n",a+1);
                        else
                        printf("Address assigned to LPT%d is 0x%X
                        \n",a+1,address);
                ptraddr++;
        }
        getch();
}
```

LISTING 3.4

```
#include"conio.h"
#include"dos.h"
#define PORT 0x378
void main()
{
        while(!kbhit())
        {
                outportb(PORT, ~inportb(PORT) );
                delay(1000);
        }
}
```

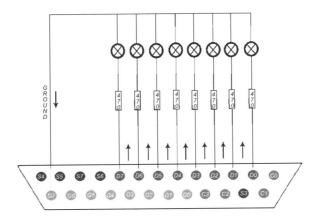

FIGURE 3.41   The circuit diagram.

We have made an electrical circuit to show you how our circuit works. It is shown in Figure 3.41. And also we have different angled pictures of the complete circuit in Figure 3.42.

Ok then, let's find out what we have to supply:

■ 1 or 2 meter parallel port cable (3 meter is acceptable but the voltage drops from 5 V to 4.7 V).

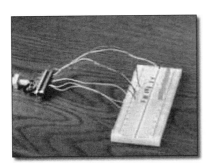

FIGURE 3.42   Pictures of the complete circuit.

- 9 assembling cables (8 go to resistance and 1 go to ground).
- A breadboard (white one in the picture) or you can solder the cables but with a breadboard you don't have to solder the cables.
- 8 LEDs (2.5 V).
- 8 resistances (470 ohm) (for not to make the LEDs garbage because of +5 V).
- A multimeter (not needed but if something wrong happens you can check the wiring with this).
- Program to make your circuit live.

Interfacing the LCD Module to a Parallel Port

You have seen LCD modules used in many electronic devices like coin phones, billing machines, and weighing machines. It is a powerful display option for stand-alone systems. Because of low power dissipation, high readability, and flexibility for programmers; LCD modules are becoming popular. In this part, we will learn how to connect an LCD module to a PC parallel port and we will prepare some library routines for LCD interfacing.

Before starting our study, let us see why you need to interface an LCD, or Liquid Crystal Display, module to the parallel port.

- If you need to modify the code, you need not have to disconnect the circuit or reprogram the chip as you do in the case of a microcontroller.
- You need to spend less: one LCD module, D-25 female connector, one potentiometer (optional), and some wires—this is what you need along with a computer.
- When you are using a computer in full-screen mode like games, movies or TV; you need to exit the application to get small updating information from the computer, i.e., if you need to watch time in that time, you have to close the games. But instead of that you can use an LCD module to display real time from the PC and you can use it along with your application. Real-time implementation from the system clock example is explained in this article. If you are good at programming, you can even connect to the Internet to get news, stock exchange updates, and make them flash in the LCD module, only if you found it important, or you can go through it by exiting your application.

LCD modules are available in a wide range like 8x1, 8x2, 16x1, 16x2, 20x2, 20x4, and 40x4. Here we have used **16x2**—that means 2 rows of 16 characters. It is a Hitachi HD44780 compatible module, having 16 pins including 2 pins for backlight.

Table 3.8 gives the pin structure of an LCD module. LCD modules without backlight will have only 14 pins. If you are using such LCDs, simply ignore the 15th and 16th pins.

TABLE 3.8	Pin Description of a HitachiHD44780 LCD	
Pin No	Symbol	Details
1	GND	Ground
2	Vcc	Supply Voltage +5V
3	Vo	Contrast adjustment
4	RS	0->Control input, 1-> Data input
5	R/W	Read/Write
6	E	Enable
7 to 14	D0 to D7	Data
15	VB1	Backlight +5V
16	VB0	Backlight ground

To program the LCD module, first we have to initialize the LCD by sending some control words. RS should be low and E should be high when we send control. R/W pin 0 means write data or control to LCD and R/W pin 1 means read data from the LCD. To send data to an LCD, make RS high, R/W low, place the data in pins 7 to 14, and make pin E high and low once. You can understand the exact method after seeing the code, later in this Chapter. To make this let us first build a circuit.

Here, we are going to write on the LCD module and not read back. So, R/W is connected to the ground directly. We need not have to input any data through, so all output pins are used in our application. Data pins of the LCD are connected to data pins of the port. Strobe signal (pin 1 of D-25 connector) is given to E (pin 6 of the LCD). Select printer (pin 17 of D-25) is connected to RS (pin 4 of the LCD).

In Figure 3.43, the LCD module is connected to the lpt port using a D-25 male connector. Pin number 3 of the LCD is for adjusting the contrast, con-

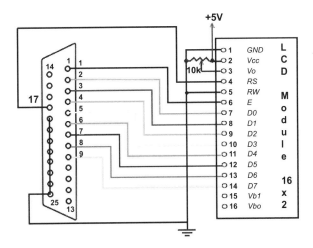

FIGURE 3.43   Connection diagram.

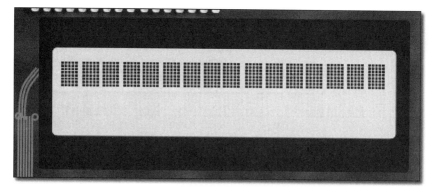

FIGURE 3.44   **An LCD in the on position.**

nected in such a way that it can be varied from 0 V to 5 V. Keep it to 0 initially.

If everything is OK, you should get the LCD module as in Figure 3.44 when the power is switched on.

If you get this screen, then we can start programming (See Fig. 3.44). Otherwise check your connections, try by varying the 10K potentiometer. If you get this display also, you can get maximum clearness by varying the pot. Here, pot was needed to be nearly 0 V. So, it is OK if you don't use pot, just connect pin 3 to the ground.

Table 3.9 explains how to write control words. When RS=0 and R/W=0, data in the pins D0 to D7 will have the following meaning.

We have left other instructions related to the read and write LCD RAM area; we will see them later. Using this information, we will write some routines for basic functions of LCDs. Now look at our first program below. Here we have written functions for all our needs in LCD interfacing. So, in our next program, we are going to change our "main" function only. You can save these functions as a library and include them in your next programs if you want.

```
#include <dos.h>
#include <string.h>
#include <conio.h>
#include <time.h>

#define PORTADDRESS 0x378 /* Enter Your Port Address Here */

#define DATA PORTADDRESS+0
#define STATUS PORTADDRESS+1
#define CONTROL PORTADDRESS+2
```

Instruction	D7	D6	D5	D4	D3	D2	D1	D0	Description
TABLE 3.9									
Clear display	0	0	0	0	0	0	0	1	Clears display and returns cursor to home position.
Cursor home	0	0	0	0	0	0	1	X	Returns cursor to home position. Also returns display being shifted to the original position.
Entry mode set	0	0	0	0	0	1	I/D	S	I/D = 0 —> cursor is in decrement position. I/D = 1 —> cursor is in increment position. S = 0 —> Shift is invisible. S = 1 —> Shift is visible.
Display ON-OFF control	0	0	0	0	1	D	C	B	D- Display, C- cursor, B- Blinking cursor = 0 —> OFF =1 —> ON
Cursor/Display shift	0	0	0	1	S/C	R/L	X	X	S/C = 0 —> Move cursor. S/C = 1 —> Shift display. R/L = 0 —> Shift left. R/L = 1 —> Shift right
Function set	0	0	1	DL	N	F	X	X	DL = 0 —> 4-bit interface. DL = 1 —> 8-bit interface. N = 0 —> 1/8 or 1/11 Duty (1 line). N = 1 —> 1/16 Duty (2 lines). F = 0 —> 5x7 dots. F = 1 —> 5x10 dots.

```
void lcd_init(void);
void lcd_write(char char2write);
void lcd_putch(char char2write);
void lcd_puts(char * str2write);
void lcd_goto(int row, int column);
void lcd_clear(void);
void lcd_home(void);
void lcd_cursor(int cursor);
void lcd_entry_mode(int mode);

void main(void)
{
    lcd_init();
    lcd_goto(1,1);
    lcd_puts("Welcome To");
```

```
    lcd_goto(1,0);
    lcd_puts("Appin Knowledge Solutions");

    while(!kbhit() ) //wait until a key is pressed...
    {}
}

void lcd_init()
{
    outportb(CONTROL, inportb(CONTROL) & 0xDF);  //config data pins as
output

    outportb(CONTROL, inportb(CONTROL) | 0x08);
    //RS is made high: control (register select)

    lcd_write(0x0f);
    delay(20);
    lcd_write( 0x01);
    delay(20);
    lcd_write( 0x38);
    delay(20);
}

void lcd_write(char char2write)
{
    outportb(DATA, char2write);
    outportb(CONTROL,inportb(CONTROL) | 0x01); /* Set Strobe */
    delay(2);
    outportb(CONTROL,inportb(CONTROL) & 0xFE);/* Reset Strobe */
    delay(2);
}

void lcd_putch(char char2write)
{
    outportb(CONTROL, inportb(CONTROL) & 0xF7);
    //RS=low: data
    lcd_write(char2write);
}

void lcd_puts(char *str2write)
{
    outportb(CONTROL, inportb(CONTROL) & 0xF7);
    //RS=low: datawhile(*str2write)
    lcd_write(*(str2write++));
}
```

```
void lcd_goto(int row, int column)
{
    outportb(CONTROL, inportb(CONTROL) | 0x08);
    if(row==2) column+=0x40;/*
    Add these if you are using LCD module with 4 columnsif(row==2)
    column+=0x14;
    if(row==3) column+=0x54;
    */lcd_write(0x80 | column);
}

void lcd_clear()
{
    outportb(CONTROL, inportb(CONTROL) | 0x08);
    lcd_write(0x01);
}

void lcd_home()
{
    outportb(CONTROL, inportb(CONTROL) | 0x08);
    lcd_write(0x02);
}

void lcd_entry_mode(int mode)
{
    /*
    if you dont call this function, entry mode sets to 2 by
    default.mode: 0 - cursor left shift, no text shift
    1 - no cursor shift, text right shift
    2 - cursor right shift, no text shift
    3 - no cursor shift, text left shift
    */
    outportb(CONTROL, inportb(CONTROL) | 0x08);
    lcd_write(0x04 + (mode%4));
}

void lcd_cursor(int cursor)
{
    /*
    set cursor:    0 - no cursor, no blink
    1 - only blink, no cursor
    2 - only cursor, no blink
    3 - both cursor and blink
    */

    outportb( CONTROL, inportb(CONTROL) | 0x08 );
```

```
        lcd_write( 0x0c + (cursor%4));
    }
```

We need not give details to all the functions above. You can understand them yourself. So, try using all the functions. In the next examples, we will generate a program that displays the system time in the LCD module. It may not have much use in DOS, but if you transfer the same to Windows, you will gain many benefits. Also, if your computer will be working in DOS most of the time, you can think of writing a TSR for the same.

In order to program to display date and time in an LCD module just replace the 'main' of the previous program with the following and run.

```
void main(void)
{
    struct time t;
    struct date d;
    char strtime[17];
    textbackground(0);
    clrscr();
    textcolor(0);
    textbackground(10);
    gotoxy(8,5);
    cputs(" ");
    gotoxy(8,4);
    cputs(" ");
    lcd_init();
    lcd_cursor(0);
    while(!kbhit())
    {
            gettime(&t);
            getdate(&d);
            lcd_goto(0,4);
            sprintf(strtime,"%02d:%02d:%02d", t.ti_hour%12, t.ti_min,
            t.ti_sec);
            lcd_puts(strtime);
            gotoxy(12,4);
            cputs(strtime);

            lcd_goto(1,3);
            sprintf(strtime,"%02d:%02d:%4d", d.da_day, d.da_mon,
            d.da_year);
            lcd_puts(strtime);
            gotoxy(11,5);
            cputs(strtime);
```

```
            delay(200);
    }
    textbackground(0);
    textcolor(7);
}
```

3.6   SERIAL COMMUNICATION: RS-232

RS-232 is the most known serial port used in transmitting the data in communication and interface. Even though a serial port is harder to program than the parallel port, this is the most effective method in which the data transmission requires fewer wires and less cost. The RS-232 is the communication line which enables data transmission by only using three wire links. The three links provide 'transmit,' 'receive,' and common ground.

The 'transmit' and 'receive' line on this connecter send and receive data between the computers. As the name indicates, the data is transmitted serially. The two pins are TXD & RXD. There are other lines on this port such as RTS, CTS, DSR, DTR, and RTS, RI. The '1' and '0' are the data which defines a voltage level of 3 V to 25 V and -3 V to -25 V respectively.

The electrical characteristics of the serial port as per the EIA (Electronics Industry Association) RS-232C Standard specifies a maximum baud rate of 20,000bps, which is slow compared to today's standard speed. For this reason, we have chosen the new RS-232D Standard, which was recently released.

The RS-232D has existed in two types, i.e., D-Type 25-pin connector and D-Type 9-pin connector, which are male connectors on the back of the PC. You need a female connector on your communication from host to guest computer. The pin outs of both D-9 & D-25 are shown in Table 3.10.

About DTE and DCE

Devices, which use serial cables for their communication, are split into two categories. These are DCE (Data Communications Equipment) and DTE (Data Terminal Equipment). Data Communication Equipments are devices such as your modem, TA adapter, plotter, etc., while Data Terminal Equipment is your computer or terminal. A typical Data Terminal Device is a computer and a typical Data Communications Device is a modem. Often people will talk about DTE to DCE or DCE to DCE speeds. DTE to DCE is the speed between your modem and computer, sometimes referred to as your terminal speed. This should run at faster speeds than the DCE to DCE speed. DCE to DCE is the link between modems, sometimes called the line speed.

TABLE 3.10

D-type 9-pin no.	D-type 25-pin no.	Pin outs	Function
3	2	RD	Receive data (serial data input)
2	3	TD	Transmit data (serial data output)
7	4	RTS	Request to send (acknowledge to modem that UART is ready to exchange data)
8	5	CTS	Clear to send (i.e., modem is ready to exchange data)
6	6	DSR	Data ready state (UART establishes a link)
5	7	SG	Signal ground
1	8	DCD	Data carrier detect (this line is active when modem detects a carrier)
4	20	DTR	Data terminal ready
9	22	RI	Ring Indicator (becomes active when modem detects ringing signal from PSTN)

Most people today will have 28.8K or 33.6K modems. Therefore, we should expect the DCE to DCE speed to be either 28.8K or 33.6K. Considering the high speed of the modem we should expect the DTE to DCE speed to be about 115,200 BPS (maximum speed of the 16550a UART). The communications program, which we use, has settings for DCE to DTE speeds. However, the speed is 9.6 KBPS, 144 KBPS, etc., and the modem speed.

If we were transferring that text file at 28.8K (DCE to DCE), then when the modem compresses it you are actually transferring 115.2 KBPS between computers and thus have a DCE to DTE speed of 115.2 KBPS. Thus, this is why the DCE to DTE should be much higher than the modem's connection speed. Therefore, if our DTE to DCE speed is several times faster than our DCE to DCE speed the PC can send data to your modem at 115,200 BPS.

Null Modem

A null modem is used to connect two DTEs together. This is used to transfer files between the computers using protocols like zmodem protocol, xmodem protocol, etc.

Figure 3.45 shows the wiring of the null modem. The main feature indicated here is to make the computer chat with the modem rather than another computer. The guest and host computer are connected through the TD, RD, and SG pins. Any data that is transmitted through the TD line from the host to guest is received on the RD line. The guest computer must have the same setup as the host. The Signal Ground (SG) line of the both must be shorted so that grounds are common to each computer.

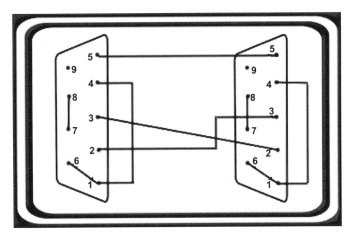

FIGURE 3.45   **Above shows the connections of the null modem using an RS 232D connector.**

The Data Terminal Ready (DTR) is looped back to Data Set Ready and Carrier Detect on both computers. When the Data Terminal Ready is asserted active, then the Data Set Ready and Carrier Detect immediately become active. At this point, the computer thinks the virtual modem to which it is connected, is ready, and has detected the carrier of the other modem.

All that's left to worry about now is the Request to Send and Clear to Send. As both computers communicate together at the same speed, flow control is not needed; thus these two lines are also linked together on each computer. When the computer wishes to send data, it asserts the Request to Send high and as it is hooked together with the Clear to Send, it immediately gets a reply that it is OK to send and does so.

The ring indicator line is only used to tell the computer that there is a ringing signal on the phone line. As we do not have a modem connected to the phone line this is left disconnected.

To know about the RS-232 ports available in your computer, right-click on "My Computer," go to 'Properties,' select the tab 'Device Manager,' go to Ports (COM & LPT). In that you will find 'Communication Port(Com1),' etc. If you right-click on that and go to properties, you will get device status. Make sure that you have enabled the port (use this port selected).

Programming the Serial Port using C/C++

There are two popular methods of sending data to or from the serial port in Turbo C. One is using outportb(PORT_ID, DATA) or outport(PORT_ID,DATA) defined in "dos.h." Another method is using the bioscom() function defined in "bios.h."

Using outportb()

The function outportb() sends a data byte to the port 'PORT_ID.' The function outport() sends a data word. These functions can be used for any port including serial ports, and parallel ports. Similarly, these are used to receive data.

- inport reads a word from a hardware port
- inportb reads a byte from a hardware port
- outport outputs a word to a hardware port
- outportb outputs a byte to a hardware port

Declaration

- int inport(int portid);
- unsigned char inportb(int portid);
- void outport(int portid, int value);
- void outportb(int portid, unsigned char value);

Remarks

- inport works just like the 80x86 instruction IN. It reads the low byte of a word from portid, the high byte from portid + 2.
- inportb is a macro that reads a byte.
- outport works just like the 80x86 instructions OUT. It writes the low byte of a value to portid, the high byte to portid + 1.
- outportb is a macro that writes the value argument.

portid

- Inport — port that inport and inportb read from;
- Outport — port that outport and outportb write to

value

- Word that outport writes to portid;
- Byte that outportb writes to portid.

If you call inportb or outportb when dos.h has been included, they are treated as macros that expand to inline code.

If you don't include dos.h, or if you do include dos.h and #undef the macro(s), you get the function(s) of the same name.

Return Value

\# inport and inportb return the value read
\# outport and outportb do not return

Using Bioscom

The macro bioscom() and function _bios_serialcom() are used in this method in the serial communication using an RS-232 connecter. First we have to set the port with the settings depending on our need and availability. In this method, the same function is used to make the settings using a control word to send data to the port and check the status of the port. These actions are distinguished using the first parameter of the function. Along with that we are sending data and the port to be used to communicate.

Here are the details of the Turbo C functions for communication ports.

Declaration

- bioscom(int cmd, char abyte, int port)
- _bios_serialcom(int cmd ,int port, char abyte)
- bioscom() and _bios_serialcom() use the bios interrupt 0x14 to perform various serial communications over the I/O ports given in port.
- cmd: The I/O operation to be performed.

portid

Port to which data is sent or from which data is read.

    0:COM1
    1:COM2
    2:COM3

a byte

When cmd = 2 or 3 (_COM_SEND or _COM_RECEIVE) parameter a byte is ignored.

When cmd = 0 (_COM_INIT), a byte is an OR combination of the following bits (one from each group). For example, if a byte = 0x8B = (0x80 | 0x08 | 0x00 | 0x03) = (_COM_1200 | _COM_ODDPARITY | _COM_STOP1 | _COM_CHR8). The communications port is set to:

    1200 baud (0x80 = _COM_1200)
    Odd parity (0x08 = _COM_ODDPARITY)

TABLE 3.11		
cmd (boiscom)	cmd(_bios_serialcom)	Action
0	_COM_INIT	Initialize the parameters to the port
1	_COM_SEND	Send the character to the port
2	_COM_RECEIVE	Receive character from the port
3	_COM_STATUS	Returns the current status of the communication port

TABLE 3.12		
value of abyte		Meaning
Bioscom	_bios_serialcom	
0x02	_COM_CHR7	7 data bits
0x03	_COM_CHR8	8 data bits
0x00	_COM_STOP1	1 stop bit
0x04	_COM_STOP2	2 stop bits
0x00	_COM_NOPARITY	No parity
0x08	_COM_ODDPARITY	Odd parity
0X10	_COM_EVENPARITY	Even parity
0x00	_COM_110	110 baud
0x20	_COM_150	150 baud
0x40	_COM_300	300 baud
0x60	_COM_600	600 baud
0x80	_COM_1200	1200 baud
0xA0	_COM_2400	2400 baud
0xC0	_COM_4800	4800 baud
0xE0	_COM_9600	9600 baud

1 stop bit (0x00 = _COM_STOP1)
8 data bits (0x03 = _COM_CHR8)

To initialize the port with the above settings we have to write, bioscom(0, 0x8B, 0). To send a data to COM1, the format of the function will be bioscom(1, data, 0). Similarly, bioscom(1, 0, 0) will read a data byte from the port.

The following example illustrates how to serial port programs. When data is available in the port, it inputs the data and displays it onto the screen and if a key is pressed the ASCII value will be sent to the port.

```
#include <bios.h>
#include <conio.h>
#define COM1      0
#define DATA_READY 0x100
#define SETTINGS ( 0x80 | 0x02 | 0x00 | 0x00)
int main(void)
{
    int in, out, status;
    bioscom(0, SETTINGS, COM1); /*initialize the port*/
    cprintf("Data sent to you:  ");
```

```
    while (1)
    {
            status = bioscom(3, 0, COM1); /*wait until get a data*/
            if (status & DATA_READY)
                    if ((out = bioscom(2, 0, COM1) & 0x7F) != 0)   /*in-
                        a data*/
                        putch(out);
                    if (kbhit())
                    {
                            if ((in = getch()) == 27)   /* ASCII of Esc*
                                    break;
                            bioscom(1, in, COM1); /*output a data*/
                    }
    }
    return 0;
}
```

When you compile and run the above program in both computers, the characters typed in one computer should appear on the other computer screen and vice versa. Initially, we set the port to the desired settings as defined in macro settings. Then we wait in an idle loop until a key is pressed or a data is available on the port. If any key is pressed, then the kbhit() function returns a nonzero value. Then we send it to the com port. Similarly, if any data is available on the port, we receive it from the port and display it on the screen.

To check the port, if you have a single computer, you can use a loop-back connection as follows. This is the most commonly used method for developing communication programs. Here, data is transmitted to that port itself.

If you run the above program with the connection as in Figure 3.46, the character entered in the keyboard should be displayed on the screen. This

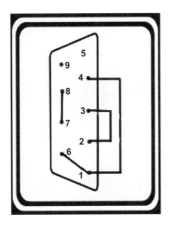

FIGURE 3.46   **Loop-back plug connection.**

method is helpful in writing serial port programs with a single computer. Also, you can make changes in the port id if your computer has 2 RS-232 ports. You can connect the com1 port to com2 of the same computer and change the port id in the program. The data sent to the com1 port should come to the com2 port. Then whatever you also type on the keyboard should appear on the screen.

3.7   USING THE MICROCONTROLLER

Before one uses the microcontroller, he or she should be well aware of the connections of the microcontroller, like which pins are to be used as data pins and which pin is to be given the power supply. Every pin of the microcontroller has a specific function and a wrong connection can completely destroy the chip. The AT89C51 microcontrollers have, in total, 40 pins. Here we will be demonstrating the pin configuration of the 8051 microcontroller. It has an on-chip RAM capacity of 128 bytes and ROM size is 4k. But the total ROM size with which it can work is 64k. So when we have programs of size more than 4k, external memory is added to it as per our requirement. If the requirement is of 8k, we can go for an external ROM of size 4k such that it makes a total ROM size of 8k (4k internal + 4k external). In this way we can go for 8k, 16k, 32k, and 64k ROM sizes.

We have different commands for using these memories. Accordingly some of the connections are to be changed. For example, if our program size is below 4k then the 31st pin, which is for external access, is to be given a power supply of 5 volts (an active low pin) but while using external memory we have to ground this pin. We can also use both internal as well as external memories simultaneously, in that case the pin is to be given a power supply of 5 volts because the microcontroller firstly reads internal ROM then goes for external ones.

8051 microcontrollers do have an internal oscillator but to run it, it needs external clock pulses so an external oscillator is added between pins 18 (XTAL1) and 19 (XTAL2). It also requires two capacitors of 30 pF value. One side of each capacitor is grounded.

It must be noted that there are various speeds of the 8051 family. Speed refers to the maximum oscillator frequency connected to the XTAL. For example, a 12 MHz chip must be connected to a crystal with a 12 MHz frequency or less. Likewise, a 20 MHz microcontroller requires a crystal frequency of no more than 20 MHz. When the 8051 is connected to a crystal oscillator and is powered up, we can observe the frequency on the XTAL2 pin using the oscilloscope.

Pin 40 of the microcontroller is given a power supply of +5 volts using a single stranded wire; this provides supply voltage to the chip. Pin 20 of the microcontroller is grounded.

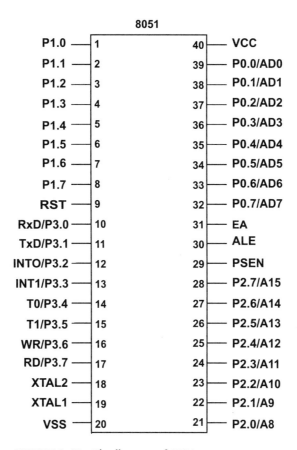

FIGURE 3.47   **Pin diagram of 8051.**

Figure 3.47, given, above shows the pin diagram of 8051.

There are four ports in 8051 and every port has 8 pins, so in total there are 32 I/O pins. There are 16 address pins and 8 data pins in it. On port 0 data and address pins are multiplexed. ALE (pin 30) indicates if P0 has address or data. When ALE = 0, it provides data to D0–D7, but when ALE =1, it has addresses at A0–A7. Therefore, ALE is used for demultiplexing addresses and data with the help of a 74LS373 latch. The ninth pin is a reset pin and it is an input and active high (normally low). Upon applying a high pulse to this pin, the microcontroller will reset and terminate all activities. This is often referred to as a power-on reset. Activating a power-on reset will cause all values in the register to be lost.

Figure 3.48 shows a complete circuit diagram of 8051.

Now we can have a program to glow LEDs. When this program is burned in the microcontroller, the LEDs start glowing as per the program written. So

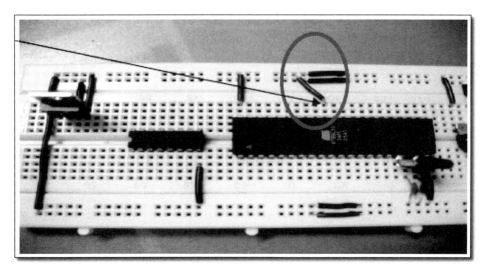

FIGURE 3.48 **Complete connections of 8051.**

as to burn the program it can first be written on a simulator like Keil C and its accuracy can be checked by debugging it there. Now the program can be burned on the microcontroller with the help of a programmer.

After the microcontroller is loaded with the program, it can be kept in the proper circuit with accurate connections and LEDs start glowing.

3.8   ACTUATORS

An **actuator** is the mechanism by which an agent acts upon an environment. The agent can be either an artificial intelligent agent or any other autonomous being.

The common forms of actuators are pneumatic, hydraulic, or electric solenoids or motors.

Pneumatic actuators: A simplified diagram of a pneumatic actuator is shown in Figure 3.49 below. It operates by a combination of force created by air and spring force. The actuator positions control the valve by transmitting its motion through the stem.

A rubber diaphragm separates the actuator housing into two air chambers. The upper chamber receives a supply of air through an opening in the top of the housing.

The bottom chamber contains a spring that forces the diaphragm against mechanical stops in the upper chamber. Finally, a local indicator is connected to the stem to indicate the position of the valve.

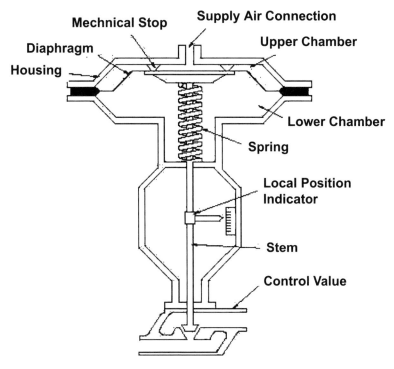

FIGURE 3.49   Pneumatic actuator: air-to-close/spring-to-open.

The position of the valve is controlled by varying the air supply pressure in the upper chamber. This results in a varying force on the top of the diaphragm. Initially, with no air supply, the spring forces the diaphragm upward against the mechanical stops and holds the valve fully open. As air supply pressure is increased from zero, its force on top of the diaphragm begins to overcome the opposing force of the spring. This causes the diaphragm to move downward and the control valve to close.

With increasing air supply pressure, the diaphragm will continue to move downward and compress the spring until the control valve is fully closed. Conversely, if air supply pressure is decreased, the spring will begin to force the diaphragm upward and open the control valve. Additionally, if supply pressure is held constant at some value between zero and maximum, the valve will position at an intermediate position. Therefore, the valve can be positioned anywhere between fully open and fully closed in response to changes in air supply pressure.

A positioner is a device that regulates the air supply pressure to a pneumatic actuator. It does this by comparing the actuator's demanded position with the control valve's actual position. The demanded position is transmitted by a pneumatic or electrical control signal from a controller to the positioner. The pneumatic actuator in Figure 3.49 is shown in Figure 3.50 with a controller and positioner added.

The controller generates an output signal that represents the demanded position. This signal is sent to the positioner. Externally, the positioner consists of an input connection for the control signal, an air supply input connection, an air supply output connection, an air supply vent connection, and a feedback linkage. Internally, it contains an intricate network of electrical transducers, airlines, valves, linkages, and necessary adjustments. Other petitioners may also provide controls for local valve positioning and gauges to indicate air supply pressure and air control pressure (for pneumatic controllers). From an operator's viewpoint, a description of the complex internal workings of a positioner is not needed. Therefore, this discussion will be limited to inputs to and outputs from the positioner.

In Figure 3.50, the controller responds to a deviation of a controlled variable from the setpoint and varies the control output signal accordingly to correct the

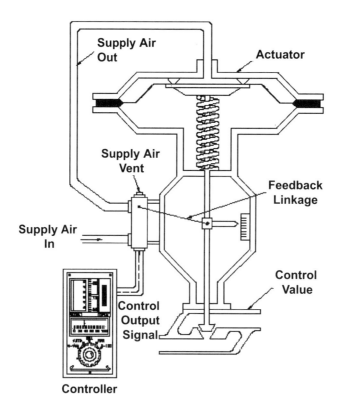

FIGURE 3.50   Pneumatic actuator with controller and positioner.

deviation. The control output signal is sent to the positioner, which responds by increasing or decreasing the air supply to the actuator. Positioning of the actuator and control valve is fed back to the positioner through the feedback linkage. When the valve has reached the position demanded by the controller, the positioner stops the change in air supply pressure and holds the valve at the new position. This, in turn, corrects the controlled variable's deviation from the setpoint.

For example, as the control signal increases, a valve inside the positioner admits more air supply to the actuator. As a result, the control valve moves downward. The linkage transmits the valve position information back to the positioner. This forms a small internal feedback loop for the actuator. When the valve reaches the position that correlates to the control signal, the linkage stops air supply flow to the actuator. This causes the actuator to stop. On the other hand, if the control signal decreases, another valve inside the positioner opens and allows the air supply pressure to decrease by venting the air supply. This causes the valve to move upward and open.

When the valve has opened to the proper position, the positioner stops venting air from the actuator and stops movement of the control valve.

An important safety feature is provided by the spring in an actuator. It can be designed to position a control valve in a safe position if a loss of air supply occurs. On a loss of air supply, the actuator in Figure 3.50 will fail to open. This type of arrangement is referred to as "air-to-close, spring-to-open" or simply "fail-open." Some valves fail in the closed position. This type of actuator is referred to as "air-to-open, spring-to-close" or "fail-closed." This "fail-safe" concept is an important consideration in nuclear facility design.

Hydraulic actuators: Pneumatic actuators are normally used to control processes requiring quick and accurate responses, as they do not require a large amount of motive force. However, when a large amount of force is required to operate a valve (for example, the main steam system valves), hydraulic actuators are normally used. Although hydraulic actuators come in many designs, piston types are the most common.

A typical piston-type hydraulic actuator is shown in Figure 3.51. It consists of a cylinder, piston, spring, hydraulic supply and returns line, and stem. The piston slides vertically inside the cylinder and separates the cylinder into two chambers. The upper chamber contains the spring and the lower chamber contains hydraulic oil.

The hydraulic supply and return line is connected to the lower chamber and allows hydraulic fluid to flow to and from the lower chamber of the actuator. The stem transmits the motion of the piston to a valve.

Initially, with no hydraulic fluid pressure, the spring force holds the valve in the closed position. As fluid enters the lower chamber, pressure in the chamber increases. This pressure results in a force on the bottom of the piston opposite to the force caused by the spring. When the hydraulic force is greater than the

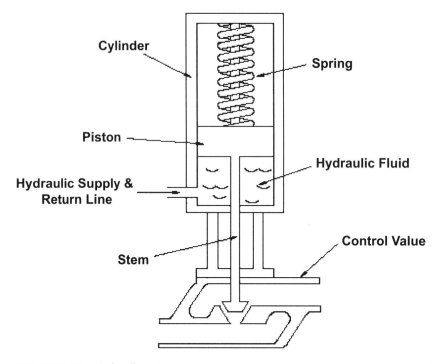

Cylinder

Spring

Piston

Hydraulic Fluid

Hydraulic Supply & Return Line

Control Value

Stem

FIGURE 3.51   **Hydraulic actuator.**

spring force, the piston begins to move upward, the spring compresses, and the valve begins to open. As the hydraulic pressure increases, the valve continues to open. Conversely, as hydraulic oil is drained from the cylinder, the hydraulic force becomes less than the spring force, the piston moves downward, and the valve closes. By regulating the amount of oil supplied or drained from the actuator, the valve can be positioned between fully open and fully closed.

The principles of operation of a hydraulic actuator are like those of the pneumatic actuator. Each uses some motive force to overcome spring force to move the valve. Also, hydraulic actuators can be designed to fail-open or fail-closed to provide a fail-safe feature.

Electric solenoid actuators: A typical electric solenoid actuator is shown in Figure 3.52. It consists of a coil, armature, spring, and stem. The coil is connected to an external current supply. The spring rests on the armature to force it downward. The armature moves vertically inside the coil and transmits its motion through the stem to the valve. When current flows through the coil, a magnetic field forms around the coil. The magnetic field attracts the armature toward the center of the coil. As the armature moves upward, the spring collapses and the valve opens. When the circuit is opened and current stops flowing to the coil, the magnetic field collapses. This allows the spring to expand and shut the valve.

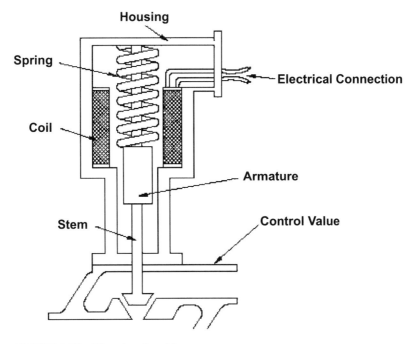

FIGURE 3.52   **Electric solenoid actuator.**

A major advantage of solenoid actuators is their quick operation. Also, they are much easier to install than pneumatic or hydraulic actuators. However, solenoid actuators have two disadvantages. First, they have only two positions: fully open and fully closed. Second, they don't produce much force, so they usually only operate relatively small valves.

Motors

This chapter introduces several types of motors commonly used in robotic and related applications.

3.8.1   DC Motors

DC motors are widely used in robotics because of their small size and high energy output. They are excellent for powering the drive wheels of a mobile robot as well as powering other mechanical assemblies.

Ratings and Specifications

Several characteristics are important in selecting a DC motor. The first two are its input ratings that specify the electrical characteristics of the motor.

Operating Voltage

If batteries are the source of power for the motor, low operating voltages are desirable because fewer cells are needed to obtain the specified voltage. However, the electronics to drive motors are typically more efficient at higher voltages. Typical DC motors may operate on as few as 1.5 volts or up to 100 volts or more. Roboticists often use motors that operate on 6, 12, or 24 volts because most robots are battery powered, and batteries are typically available with these values.

Operating Current

The ideal motor would produce a great deal of power while requiring a minimum of current. However, the current rating (in conjunction with the voltage rating) is usually a good indication of the power output capacity of a motor. The power input (current times voltage) is a good indicator of the mechanical power output. Also, a given motor draws more current as it delivers more output torque. Thus, current ratings are often given when the motor is stalled. At this point it is drawing the maximum amount of current and applying maximum torque. A low-voltage (e.g., 12 volts or less) DC motor may draw from 100 mA to several amperes at stall, depending on its design.

The next three ratings describe the motor's output characteristics.

Speed

Usually, this is specified as the speed in rotations per minute (RPM) of the motor when it is unloaded, or running freely, at its specified operating voltage. Typical DC motors run at speeds from one to twenty thousand RPM. Motor speed can be measured easily by mounting a disk or LEGO pulley wheel with one hole on the motor, and using a slotted optical switch and oscilloscope to measure the time between the switch openings.

Torque

The torque of a motor is the rotary force produced on its output shaft. When a motor is stalled it is producing the maximum amount of torque that it can produce. Hence the torque rating is usually taken when the motor has stalled and is called the *stall torque*. The motor torque is measured in ounces-inches (in the English system) or Newton-meters (metric). The torque of small electric motors is often given in milli-Newton-meters (mN-m) or 1/1000 of a N-m. A rating of one ounce-inch means that the motor is exerting a tangential force of one ounce at a radius of one inch from the center of its shaft. Torque ratings may vary from less than one ounce-inch to several dozen ounce-inches for large motors.

Power

The power of a motor is the product of its speed and torque. The power output is greatest at about halfway between the unloaded speed (maximum speed, no torque) and the stalled state (maximum torque, no speed). The output power in watts is about (torque) x (rpm) / 9.57.

3.8.2   Controlling a DC Motor

Speed, Torque, and Gear Reduction

It was mentioned earlier that the power delivered by a motor is the product of its speed and the torque at which the speed is applied. If one measures this power over the full range of operating speeds—from unloaded full speed to stall—one gets a bell-shaped curve of motor power output.

When unloaded, the motor is running at full speed, but at zero torque, thus producing zero power. Conversely, when stalled, the motor is producing its maximum torque output, but at zero speed—also producing zero power! Hence, the maximum power output must lie somewhere in between, at about one-half of the maximum speed and of the maximum torque.

A typical DC motor operates at speeds that are far too high to be useful, and at torques that are far too low. *Gear reduction* is the standard method by which a motor is made useful.

The motor shaft is fitted with a gear of small radius that meshes with a gear of large radius. The motor's gear must revolve several times in order to cause the large gear to revolve once (see Figure 3.53). The speed of rotation is thus decreased, but overall power is preserved (except for losses due to friction) and therefore the torque must increase. By ganging together several stages of this gear reduction, a strong torque can be produced at the final stage.

The challenge when designing a high-performance gear reduction for a competitive robot is to determine the amount of reduction that will allow the motor to operate at highest efficiency. If the normal operating speed of a motor/gear-train assembly is faster than the peak efficiency point, the gear-train will

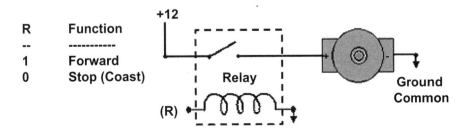

FIGURE 3.53

be able to accelerate quickly, but will not be operating at peak efficiency once it has reached the maximum velocity. Remember that the wheel is part of the drive train and gearing, and its size, the velocity desired, the motor characteristics, and other factors all affect the optimum gear ratio. While calculations can provide a guide, experimentation is necessary to determine the best gear-train.

H-bridge

You take a battery; hook the positive side to one side of your DC motor. Then you connect the negative side of the battery to the other motor lead. The motor spins forward. If you swap the battery leads the motor spins in reverse.

Ok, that's basic. Now lets say you want a Micro Controller Unit (MCU) to control the motor, how would you do it? Well, for starters you get a device that would act like a solid state switch, a transistor, and hook it up to the motor.

NOTE ▶ If you connect up these relay circuits, remember to put a diode across the coil of the relay. This will keep the spike voltage (back EMF), coming out of the coil of the relay, from getting into the MCU and damaging it. The anode, which is the arrow side of the diode, should connect to ground. The bar, which is the cathode side of the diode, should connect to the coil where the MCU connects to the relay.

If you connect this circuit to a small hobby motor you can control the motor with a processor (MCU, etc.). Applying a logical one, (+12 volts in our example) to point A causes the motor to turn forward. Applying a logical zero, (ground) causes the motor to stop turning (to coast and stop).

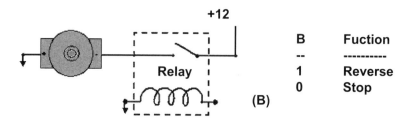

B	Fuction
--	----------
1	Reverse
0	Stop

FIGURE 3.54

Hook the motor up in this fashion and the circuit turns the motor in reverse when you apply a logical one (+12 volts) to point B. Apply a logical zero, which is usually a ground, and the motor stops spinning.

If you hook up these circuits you can only get the motor to stop or turn in one direction, forward for the first circuit or reverse for the second circuit.

Motor Speed

You can also pulse the motor control line, (A or B) on and off. This powers the motor in short bursts and gets varying degrees of torque, which usually translates into variable motor speed.

But if you want to be able to control the motor in both forward and reverse with a processor, you will need more circuitry. You will need an H-bridge. Notice the "H"-looking configuration in Figure 3.55. Relays configured in this fashion make an H-bridge. The "high side drivers" are the relays that control the positive voltage to the motor. This is called sourcing current.

The "low side drivers" are the relays that control the negative voltage to sink current to the motor. "Sinking current" is the term for connecting the circuit to the negative side of the power supply, which is usually ground.

So, you turn on the upper left and lower right circuits, and power flows through the motor forward, i.e., 1 to A, 0 to B, 0 to C, and 1 to D.

Then for reverse you turn on the upper right and lower left circuits and power flows through the motor in reverse, i.e., 0 to A, 1 to B, 1 to C, and 0 to D.

Caution: You should be careful not to turn on both circuits on one side and the other, or you have a direct short which will destroy your circuit; for example: A and C or B and D both high (logical 1).

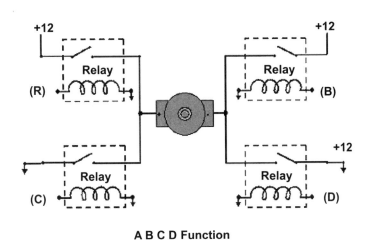

A	B	C	D	Function	
1	0	0	1	Forward	
0	1	1	0	Reverse	
1	1	0	0	Brake	
0	0	1	1	Brake	
1	0	1	0	Fuse Test	: -)
0	1	0	1	Fuse Test	: -)

Don'e de the Fuse bests

FIGURE 3.55

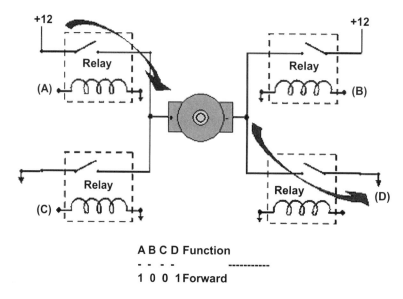

A B C D Function
- - - -   ------------
1 0 0 1 Forward

FIGURE 3.56

Semiconductor H-bridges

We can better control our motor by using transistors or Field Effect Transistors (FETs). Most of what we have discussed about the H-bridge relays is true of these circuits. You don't need diodes that were across the relay coils now. You should use diodes across your transistors though. See Figure 3.56 see how they are connected.

These solid state circuits provide power and ground connections to the motor, as did the relay circuits. The high side drivers need to be current "sources" which is what PNP transistors and P-channel FETs are good at. The low side drivers need to be current "sinks" which is what NPN transistors and N-channel FETs are good at.

If you turn on the two upper circuits, the motor resists turning, so you effectively have a breaking mechanism. The same is true if you turn on both of the lower circuits. This is because the motor is a generator and when it turns it generates a voltage. If the terminals of the motor are connected (shorted), then the voltage generated counteracts the motors freedom to turn. It is as if you are applying a similar but opposite voltage to the one generated by the motor being turned. Vis-á-vis, it acts like a brake.

To be nice to your transistors, you should add diodes to catch the back voltage that is generated by the motor's coil when the power is switched on and off. This flyback voltage can be many times higher than the supply voltage! If you don't use diodes, you could burn out your transistors.

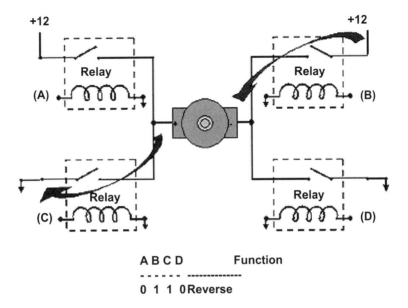

| A B C D | Function |
| - - - - - - ------------- |
| 0 1 1 0 | Reverse |

FIGURE 3.57

Transistors, being a semiconductor device, will have some resistance, which causes them to get hot when conducting much current. This is called not being able to sink or source very much power, i.e., not able to provide much current from ground or from plus voltage.

Mosfets are much more efficient, they can provide much more current and not get as hot. They usually have the flyback diodes built in so you don't need the diodes anymore. This helps guard against flyback voltage frying your MCU.

To use mosfets in an H-bridge, you need P-channel mosfets on top because they can "source" power, and N-channel mosfets on the bottom because then can "sink" power. N-channel mosfets are much cheaper than P-channel mosfets, but N-channel mosfets used to source power require about 7 volts more than the supply voltage, to turn on. As a result, some people manage to use N-channel mosfets, on top of the H-bridge, by using cleaver circuits to overcome the breakdown voltage.

It is important that the four quadrants of the H-bridge circuits be turned on and off properly. When there is a path between the positive and ground side of the H-bridge, other than through the motor, a condition exists, called "shoot through." This is basically a direct short of the power supply and can cause semi-conductors to become ballistic in circuits with large currents flowing. There are H-bridge chips available that are much easier, and safer, to use than designing your own H-bridge circuit.

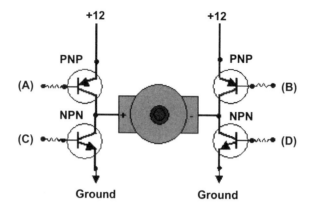

```
A B C D  Function
- - - - - -      -------------
1 0 0 1      Forward
0 1 1 0      Reverse
1 1 0 0      Brake
0 0 1 1      Brake
1 0 1 0      Fuse test  : -)
0 1 0 1      Fuse test  : -)
Don't do the fuse tests
```

FIGURE 3.58

H-bridge Devices

The L293 has 2 H-bridges, can provide about 1 amp to each and occasional peak loads to 2 amps. Motors typically controlled with this controller are near the size of a 35 mm film plastic canister.

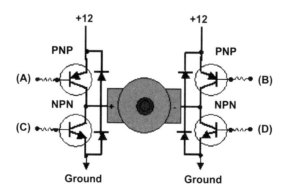

FIGURE 3.59

The L298 has 2 H-bridges on board, can handle 1 amp, and peak current draws to about 3 amps. You often see motors between the size of a 35 mm film plastic canister and a coke can, driven by this type of H-bridge. The LMD18200 has one H-bridge on board, can handle about 2 or 3 amps, and can handle a peak of about 6 amps. This H-bridge chip can usually handle an average motor about the size of a coke. There are several more commercially designed H-bridge chips as well.

There! That's the basics about motors and H-bridges! Hope it helps and be safe!

Darlington Connection

This is two transistors connected together so that the current amplified by the first is amplified further by the second transistor. The overall current gain is equal to the two individual gains multiplied together:

Darlington pair current gain, $h_{FE} = h_{FE1} \times h_{FE2}$
(h_{FE1} and h_{FE2} are the gains of the individual transistors).

This gives the Darlington pair a very high current gain, such as 10,000, so that only a tiny base current is required to make the pair switch on.

A Darlington pair behaves like a single transistor with a very high current gain. It has three leads (**B, C,** and **E**) which are equivalent to the leads of a standard individual transistor. To turn on there must be 0.7 V across both the base-emitter junctions who are connected in series inside the Darlington pair, therefore it requires 1.4 V to turn on.

(1-2) Avoid a Darlington connection. (High Gain)

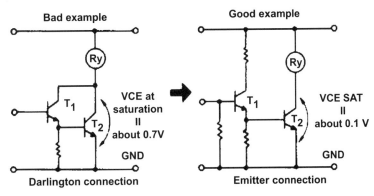

- **Because of wasted power consumption heat develops.**
- **A strong transistor for T$_1$ is required.**

- **T$_2$ conducts completely.**
- **T$_1$ is sufficient for signal use.**

FIGURE 3.60

Darlington pairs are available as complete packages but you can make up your own from two transistors; TR1 can be a low power type, but normally TR2 will need to be high power. The maximum collector current Ic(max) for the pair is the same as Ic(max) for TR2.

A Darlington pair is sufficiently sensitive to respond to the small current passed by your skin and it can be used to make a **touch-switch** as shown in Figure 3.61. For this circuit, which just lights an LED, the two transistors can be any general-purpose low-power transistors. The 100kΩ resistor protects the transistors if the contacts are linked with a piece of wire.

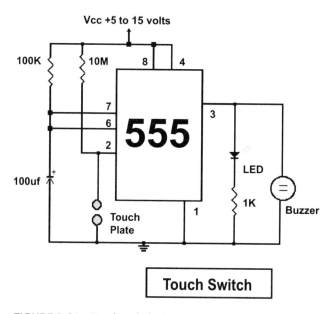

FIGURE 3.61 Touch switch circuit.

3.8.3 Pulse Width Modulation

Pulse width modulation is a technique for reducing the amount of power delivered to a DC motor. Instead of reducing the voltage operating the motor (which would reduce its power), the motor's power supply is rapidly switched on and off. The percentage of time that the power is on determines the percentage of full operating power that is accomplished. This type of motor speed control is easier to implement with digital circuitry. It is typically used in mechanical systems that will not need to be operated at full power all of the time. For an ELEC 201 robot, this would often be a system other than the main drivetrain or when the main drivetrain is steered.

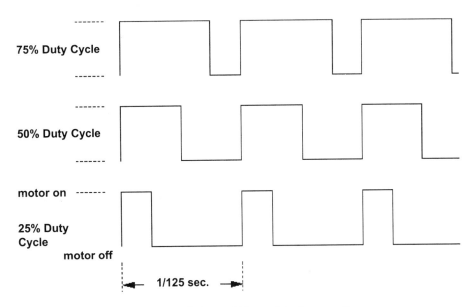

FIGURE 3.62   Example of several pluse width modulation waveforms.

Figure 3.62 illustrates this concept, showing pulse width modulation signals to operate a motor at 75%, 50%, and 25% of the full power potential.

A wide range of frequencies could be used for the pulse width modulation signal. The ELEC 201 system software used to control the motors operates at 1,000 Hertz.

A PWM waveform consisting of eight bits, each of which may be on or off, is used to control the motor. Every 1/1000 of a second, a control bit determines whether the motor is enabled or disabled. Every 1/125 of a second the waveform is repeated. Therefore, the control bit make 8 checks per cycle, meaning the PWM waveform may be adjusted to eight power levels between off and full on. This provides the RoboBoard with eight motor speeds.

3.8.4   Stepper Motors

The shaft of a stepper motor moves between discrete rotary positions typically separated by a few degrees. Because of this precise position controllability, stepper motors are excellent for applications that require high positioning accuracy. Stepper motors are used in X-Y scanners, plotters, and machine tools, floppy and hard disk drive head positioning, computer printer head positioning, and numerous other applications.

Stepper motors have several electromagnetic coils that must be powered sequentially to make the motor turn, or step, from one position, to the next. By

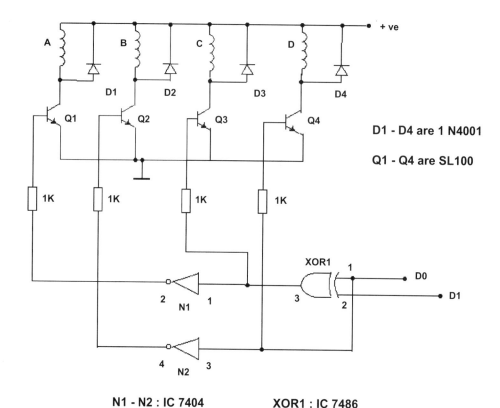

D1 - D4 are 1 N4001

Q1 - Q4 are SL100

N1 - N2 : IC 7404   XOR1 : IC 7486

PINS 14 & 7 of both the ICs are to be connected to +5 V and ground respectively.

FIGURE 3.63

reversing the order that the coils are powered, a stepper motor can be made to reverse direction. The rate at which the coils are respectively energized determines the velocity of the motor up to a physical limit. Typical stepper motors have two or four coils. Anyway, here is a very simple stepper controller shown in Figure 3.63.

How Stepper Motors Work

We've all experimented with small "hobby motors," or free-spinning DC motors. Have you ever tried to position something accurately with one? It can be pretty difficult. Even if you get the timing just right for starting and stopping the motor, the armature does not stop immediately. DC motors have very gradual acceleration and declaration curves; stabilization is slow. Adding gears to the motor will help to reduce this problem, but overshoot is still present and will throw off the anticipated stop position. The only way to effectively use a DC motor for precise positioning is to use a servo. Servos usually implement a small DC motor, a feedback mechanism (usually a potentiometer attached to the shaft by gearing or other means), and a control circuit which compares the position of the motor with the desired position, and moves the motor accordingly. This can get fairly complex and expensive for most hobby applications.

Stepper motors, however, behave differently than standard DC motors. First of all, they cannot run freely by themselves. Stepper motors do as their name suggests—they "step" a little bit at a time. Stepper motors also differ from DC motors in their torque-speed relationship. DC motors generally are not very good at producing high torque at low speeds, without the aid of a gearing mechanism. Stepper motors, on the other hand, work in the opposite manner. They produce the highest torque at low speeds. Stepper motors also have another characteristic, holding torque, which is not present in DC motors. Holding torque allows a stepper motor to hold its position firmly when not turning. This can be useful for applications where the motor may be starting and stopping, while the force acting against the motor remains present. This eliminates the need for a mechanical brake mechanism. Steppers don't simply respond to a clock signal, they have several windings which need to be energized in the correct sequence before the motor's shaft will rotate. Reversing the order of the sequence will cause the motor to rotate the other way. If the control signals are not sent in the correct order, the motor will not turn properly. It may simply buzz and not move, or it may actually turn, but in a rough or jerky manner. A circuit which is responsible for converting step and direction signals into winding energization patterns is called a translator. Most stepper motor control systems include a driver in addition to the translator, to handle the current drawn by the motor's windings.

Figure 3.64 shows a basic example of the "translator + driver" type of configuration. Notice the separate voltages for logic and for the stepper motor. Usually the motor will require a different voltage than the logic portion of the system. Typically, logic voltage is +5 Vdc and the stepper motor voltage can range

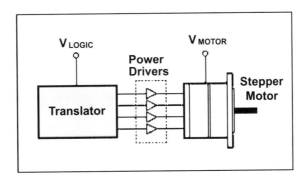

FIGURE 3.64   **A typical translator/driver connection.**

from +5 Vdc up to about +48 Vdc. The driver is also an "open collector" driver, wherein it takes its outputs to GND to activate the motor's windings. Most semi-conductor circuits are more capable of sinking (providing a GND or negative voltage) than sourcing (outputting a positive voltage).

Common Characteristics of Stepper Motors

Stepper motors are not just rated by voltage. The following elements character-ize a given stepper motor:

Voltage

Stepper motors usually have a voltage rating. This is either printed directly on the unit, or is specified in the motor's datasheet. Exceeding the rated voltage is sometimes necessary to obtain the desired torque from a given motor, but doing so may produce excessive heat and/or shorten the life of the motor.

Resistance

Resistance-per-winding is another characteristic of a stepper motor. This re-sistance will determine current draw of the motor, as well as affect the motor's torque curve and maximum operating speed.

Degrees per Step

This is often the most important factor in choosing a stepper motor for a given application. This factor specifies the number of degrees the shaft will rotate for each full step. Half-step operation of the motor will double the number of steps/revolutions, and cut the degrees-per-step in half. For unmarked motors, it is often possible to carefully count, by hand, the number of steps per revolution of the motor. The degrees per step can be calculated by dividing 360 by the number of steps in 1 complete revolution. Common degree/step numbers include: 0.72, 1.8, 3.6, 7.5, 15, and even 90. Degrees per step are often referred to as the reso-lution of the motor. As in the case of an unmarked motor, if a motor has only the number of steps/revolutions printed on it, dividing 360 by this number will yield the degree/step value.

Types of Stepper Motors

Stepper motors fall into two basic categories: permanent magnet and variable reluctance. The type of motor determines the type of drivers, and the type of translator used. Of the permanent magnet stepper motors, there are several "subflavors" available. These include the unipolar, bipolar, and multiphase va-rieties.

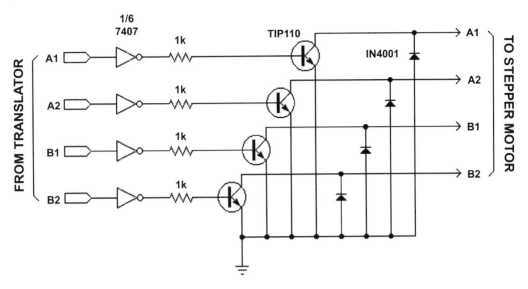

FIGURE 3.65   **A typical unipolar stepper motor driver circuit. Note the 4 back EMF protection diodes.**

Permanent Magnet Stepper Motors

Unipolar Stepper Motors

Unipolar motors are relatively easy to control. A simple 1-of-'n' counter circuit can generate the proper stepping sequence, and drivers as simple as 1 transistor per winding are possible with unipolar motors. Unipolar stepper motors are characterized by their center-tapped windings. A common wiring scheme is to take all the taps of the center-tapped windings and feed them +MV (Motor Voltage). The driver circuit would then ground each winding to energize it.

Unipolar stepper motors are recognized by their center-tapped windings. The number of phases is twice the number of coils, since each coil is divided in two. So the diagram above (Figure 3.65), which has two center-tapped coils, represents the connection of a 4-phase unipolar stepper motor.

In addition to the standard drive sequence, high-torque and half-step drive sequences are also possible. In the high-torque sequence, two windings are active at a time for each motor step. This two-winding combination yields around 1.5 times more torque than the standard sequence, but it draws twice the current. Half-stepping is achieved by combining the two sequences. First, one of the windings is activated, then two, then one, etc. This effectively doubles the number of steps the motor will advance for each revolution of the shaft, and it cuts the number of degrees per step in half.

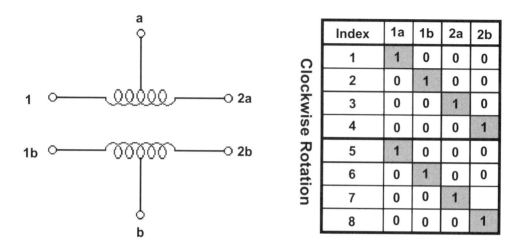

FIGURE 3.66   Unipolar stepper motor coil setup (left) and 1-phase drive pattern (right).

Index	1a	1b	2a	2b
1	1	0	0	1
2	1	1	0	0
3	0	1	1	0
4	0	0	1	1
5	1	0	0	1
6	1	1	0	0
7	0	1	1	0
8	0	0	1	1

Clockwise Rotation

Alternate Full-Step Sequence (Provides more torque)

Index	1a	1b	2a	2b
1	1	0	0	0
2	1	1	0	0
3	0	1	0	0
4	0	1	1	0
5	0	0	1	0
6	0	0	1	1
7	0	0	0	1
8	1	0	0	1
9	1	0	0	0
10	1	1	0	0
11	0	1	0	0
12	0	1	1	0
13	0	0	1	0
14	0	0	1	1
15	0	0	0	1
16	1	0	0	1

Clockwise Rotation

Half-Step Sequence

FIGURE 3.67   Two-phase stepping sequence (left) and half-step sequence (right).

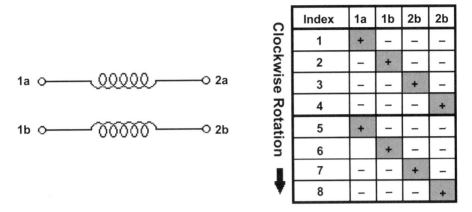

Index	1a	1b	2b	2b
1	+	–	–	–
2	–	+	–	–
3	–	–	+	–
4	–	–	–	+
5	+	–	–	–
6		+	–	–
7	–	–	+	–
8	–	–	–	+

Clockwise Rotation

FIGURE 3.68 Bipolar stepper motor coil setup (left) and drive pattern (right).

Bipolar Stepper Motors

Unlike unipolar stepper motors, bipolar units require more complex driver circuitry. Bipolar motors are known for their excellent size/torque ratio, and provide more torque for their size than unipolar motors. Bipolar motors are designed with separate coils that need to be driven in either direction (the polarity needs to be reversed during operation) for proper stepping to occur. This presents a driver challenge. Bipolar stepper motors use the same binary drive pattern as a unipolar motor, only the '0' and '1' signals correspond to the polarity of the voltage applied to the coils, not simply 'on-off' signals. Figure 3.68 shows a basic 4-phase bipolar motor's coil setup and drive sequence.

A circuit known as an "H-bridge" (Figure 3.69) is used to drive bipolar stepper motors. Each coil of the stepper motor needs its own H-bridge driver circuit. Typical bipolar steppers have 4 leads, connected to two isolated coils in the motor. ICs specifically designed to drive bipolar steppers (or DC motors) are available (popular are the L297/298 series from ST Microelectronics, and the LMD18T245 from National Semiconductor). Usually these IC modules only contain a single H-bridge circuit inside of them, so two of them are required for driving a single bipolar motor. One problem with the basic (transistor) H-bridge circuit is that with a certain combination of input values (both '1's) the result is that the power supply feeding the motor becomes shorted by the transistors. This could cause a situation where the transistors and/or power supply may be destroyed. A small XOR logic circuit was added in Figure 3.69 to keep both inputs from being seen as '1's by the transistors.

Another characteristic of H-bridge circuits is that they have electrical "brakes" that can be applied to slow or even stop the motor from spinning freely when not

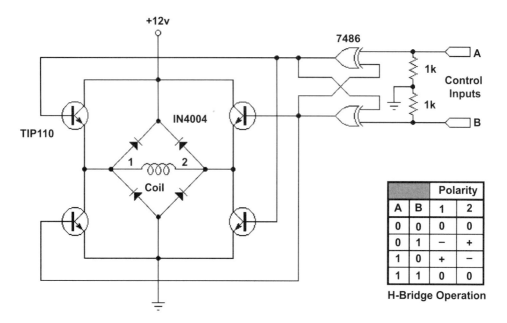

FIGURE 3.69 A typical H-bridge circuit. The 4 diodes clamp inductive kickback.

moving under control by the driver circuit. This is accomplished by essentially shorting the coil(s) of the motor together, causing any voltage produced in the coils during rotation to "fold back" on itself and make the shaft difficult to turn. The faster the shaft is made to turn, the more the electrical "brakes" tighten.

Variable Reluctance Stepper Motors

Sometimes referred to as hybrid motors, variable reluctance stepper motors are the simplest to control over other types of stepper motors. Their drive sequence

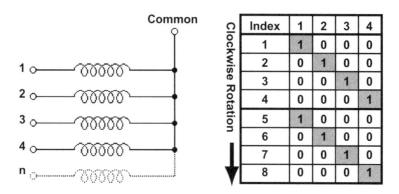

FIGURE 3.70 Variable reluctance stepper motor coil setup (left) and drive pattern (right).

is simply to energize each of the windings in order, one after the other (see drive pattern table below). This type of stepper motor will often have only one lead, which is the common lead for all the other leads. This type of motor feels like a DC motor when the shaft is spun by hand; it turns freely and you cannot feel the steps. This type of stepper motor is not permanently magnetized like its unipolar and bipolar counterparts.

Example Translator Circuits

In this section, examples of basic stepper motor translation circuits are shown. Not all of these examples have been tested, so be sure to prototype the circuit before soldering anything.

Figure 3.71 illustrates the simplest solution to generating a one-phase drive sequence. For unipolar stepper motors, the circuit in Figure 3.71, or for bipolar stepper motors, the circuit in Figure 3.72 can be connected to the 4 outputs of this circuit to provide a complete translator + driver solution. This circuit is limited in that it cannot reverse the direction of the motor. This circuit would be most useful in applications where the motor does not need to change directions.

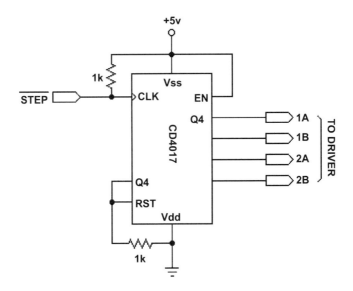

FIGURE 3.71   A simple, single-direction, single-phase drive translator.

Figure 3.72 is a translator for a two-phase operation (believed to have originated from *The Robot Builders Bonanza*) book, by Gordon McComb. We have used this circuit in the past and seem to recall that it had a problem. This may not

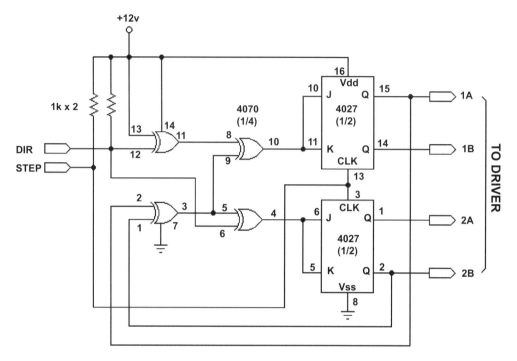

FIGURE 3.72 A simple, bidirectional, two-phase drive stepper motor translator circuit.

be the case when you reverse the direction and continue stepping, the motor will advance one more step in the previous direction it was going before responding. As always, prototype this circuit to be sure it will work for your application before you build anything with it.

There are several standard stepper motor translation circuits that use discrete logic ICs. Below you will find yet another one of these. The circuit in Figure 3.73 has not been tested, but theoretically should work without problems.

Words of Caution

When making connections to either a PC parallel port, or I/O pins of a microcontroller, be sure to isolate the motor well. High-voltage spikes of several hundred volts are possible as back EMF from the stepper motor coils. Always use clamping diodes to short these spikes back to the motor's power bus. The use of optical isolation devices (optoisolators) will add yet another layer or protection between the delicate control logic and the high-voltage potentials which may be present in the power output stage. Whenever possible, use separate power supplies for the motor and the translator/microcontroller. This further reduces the chance of destructive voltages reaching the controller, and reduces or eliminates power supply noise that may be introduced by the motor.

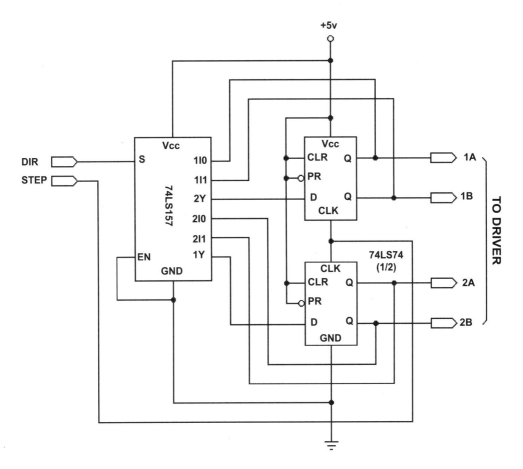

FIGURE 3.73   **Another example of a two-phase drive translator circuit, this time using a multiplexer.**

Complete Software Control

Under complete software control, there is no translator circuit external to the parallel port or microcontroller. This scheme reduces parts count, component cost, and makes for simpler board design. On the other hand, it places the responsibility of generating all of the sequencing signals on the software. If the PC or microcontroller is not fast enough (due to code inefficiency or slow processor speed), or too many motors are driven simultaneously, things can begin to slow down. Interrupts and other system events can plague the control software more in this case. Despite the downfalls of addressing a stepper motor directly in this manner, it is definitely the easiest and most straightforward approach to controlling a stepper motor. This method of controlling a motor can also be useful where the hardware is not critical at first and a simple interface is needed to allow more time to be spent on the development of the software before the hardware is refined.

3.8.5   Servo Motor

A servo motor has three wires: power, ground, and control. The power and ground wires are simply connected to a power supply. Most servo motors operate from five volts.

The control signal consists of a series of pulses that indicate the desired position of the shaft. Each pulse represents one position command. The length of a pulse in time corresponds to the angular position. Typical pulse times range from 0.7 to 2.0 milliseconds for the full range of travel of a servo shaft. Most servo shafts have a 180-degree range of rotation. The control pulse must repeat every 20 milliseconds. There are no servo motors in the present ELEC 201 kit.

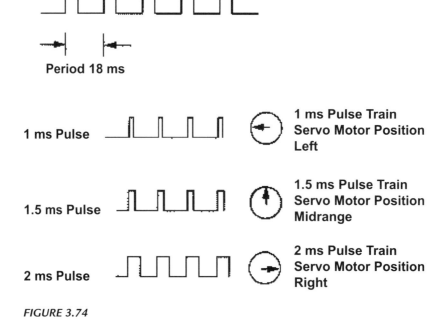

FIGURE 3.74

Servo Motors

Servo motors incorporate several components into one device package:

- a small DC motor;
- a gear reduction drive for torque increase;
- an electronic shaft position sensing and control circuit.

The output shaft of a servo motor does not rotate freely, but rather is commanded to move to a particular angular position. The electronic sensing and control circuitry—the servo feedback control loop—drives the motor to move the shaft to the commanded position. If the position is outside the range of movement of the shaft, or if the resisting torque on the shaft is too great, the motor will continue trying to attain the commanded position.

Servo motors are used in model radio-controlled airplanes and helicopters to control the position of wing flaps and other flight control mechanisms.

Servo Motor Control

A servo motor has three wires: power, ground, and control. The power and ground wires are simply connected to a power supply. Most servo motors operate from five volts.

The servo controller receives position commands through a serial connection, which can be provided by using one I/O pin of another microcontroller, or a PCs serial port! The communication protocol, that is used for this controller, is the same with the protocol of all the famous servo controllers of Scott Edwards Electronics Inc., this makes this new controller 100% compatible with all the programs that have been written for the "SSC" controllers! However, if you want to write your own software, it is as easy as sending positioning data to the serial port as follows:

> Byte1 = Sync (255)
> Byte2 = Servo #(0–15)
> Byte3 = Position (0–254)

So sending a 255, 4,150 would move servo 4 to position 150, sending 255,12,35 would move servo 12 to position 35.

The standards of the serial communication should be the following: 9600 baud, 8 data bits, 1 stop bit, and no parity.

The control signal consists of a series of pulses that indicate the desired position of the shaft. Each pulse represents one position command. The length of a pulse in time corresponds to the angular position. Typical pulse times range from 0.7 to 2.0 milliseconds for the full range of travel of a servo shaft. Most servo shafts have a 180-degree range of rotation. The control pulse must repeat every 20 milliseconds. This pulse signal will cause the shaft to locate itself at the midway position +/-90 degrees. The shaft rotation on a servo motor is limited to approximately 180 degrees (+/-90 degrees from center position). A 1 ms pulse will rotate the shaft all the way to the left, while a 2 ms pulse will turn the shaft all the way to the right. By varying the pulse width between 1 and 2 ms, the servo motor shaft can be rotated to any degree position within its range.

4 WHEELED MOBILE ROBOTS

In This Chapter

- Introduction
- Classification of Wheeled Mobile Robots (WMRs)
- Kinematics and Mathematical Modeling of WMRs
- Control of WMRs
- Simulation of WMRs Using Matlab
- The Identification and Elimination of the Problem
- Modifying the Model to Make the Variation in Delta Continuous
- Developing the Software and Hardware Model of an All-Purpose Research WMR

4.1   INTRODUCTION

Wheeled Mobile Robots (WMRs) have been an active area of research and development over the past three decades. This long-term interest has been mainly fueled by the myriad of practical applications that can be uniquely addressed by mobile robots due to their ability to work in large (potentially unstructured and hazardous) domains.

WMRs are increasingly present in industrial and service robotics, particularly when flexible motion capabilities are required on reasonably smooth grounds and surfaces. Several mobility configurations (wheel number and type, their location and actuation, single- or multibody vehicle structure) can be found in the applications.

Mobile robots have quite simple mathematical models to describe their instantaneous motion capabilities; especially compared to serial, parallel, and humanoid robots. However, this only holds for single mobile robots only, because the modeling does become complex as soon as one begins to add *trailers* to mobile robots. Airport luggage carts are good examples of such mobile robot trains.

4.2   CLASSIFICATION OF WHEELED MOBILE ROBOTS (WMRS)

The wheeled mobile robots can have a large number of possible wheel configurations and kinematic designs. Each type of configuration has its merits and demerits with respect to the application. The following is the classification of WMRs according to their wheel geometry.

4.2.1   Differentially Driven WMRs

 Differential drive configuration is the most common wheeled mobile robot configuration. It is used because of its simplicity and versatility. It is the easiest to implement and to control. A differentially driven WMR consists of two driving wheels and one or two castor wheels. In a differentially driven WMR the relative motion of the two driving wheels with respect to each other achieves the required motion. The castor wheels are used just to support the structure.

The motion of a differentially driven WMR is simple. The straight-line motion is attained in the robot when the two driving wheels rotate at the same speed. Motion in the reverse direction is achieved by the rotation of the wheels in the opposite direction. Turning is achieved by braking one wheel and rotating

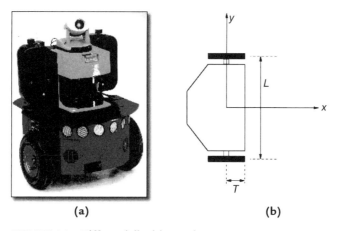

(a)   (b)

FIGURE 4.1   **Differentially driven robot.**

the other. The robot rotates about the fixed wheel. Sharp turning can be attained by rotating both the wheels in opposite directions. Motion along an arc can be attained by the differential motion of both the wheels with respect to each other.

Figure 4.1 represents a general arrangement of a differentially driven wheeled mobile robot.

4.2.2   Car-type WMRs

Car-type configurations (see Figure 4.2) employing one (tricycle-drive) or two driven front wheels and two passive rear wheels (or vice versa) are fairly common in AGV applications because of their inherent simplicity. One problem associated with the tricycle-drive configuration is that the vehicle's center of gravity tends to move away from the front wheel when traversing up an incline, causing a loss of traction. As in the case of Ackerman-steered designs, some surface damage and induced heading errors are possible when actuating the steering while the platform is not moving.

Ackerman steering provides a fairly accurate odometry solution while supporting the traction and ground clearance needs of all-terrain operation. Ackerman steering is thus the method of choice for outdoor autonomous vehicles. Associated drive implementations typically employ a gasoline or diesel engine coupled to a manual or automatic transmission, with power applied to four wheels through a transfer case, a differential, and a series of universal joints. From a military perspective, the use of existing-inventory equipment of this type simplifies some of the logistics problems associated with vehicle maintenance. In

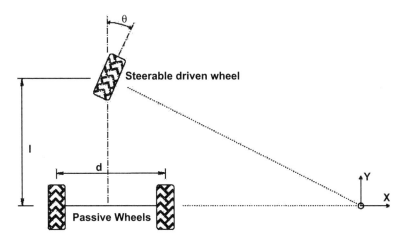

FIGURE 4.2   Tricycle-driven configurations employing a steerable driven wheel and two passive trailing wheels can derive heading information directly from a steering angle encoder or indirectly from differential odometry.

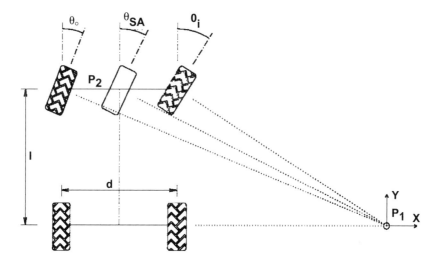

FIGURE 4.3   In an Ackerman-steered vehicle, the extended axes for all wheels intersect in a common point.

addition, reliability of the drive components is high due to the inherited stability of a proven power train. (Significant interface problems can be encountered, however, in retrofitting off-the-shelf vehicles intended for human drivers to accommodate remote or computer control.)

4.2.3   Omnidirectional WMRs

Omnidirectional movement of WMRs is of great interest for complete maneuverability. This kind of wheel configuration imposes no kinematic constraint on the robot chassis. Hence, the robot can freely change its direction at any instant. The odometry solution for this configuration is done in a similar fashion to that for differential drive, with position and velocity data derived from the motor (or wheel) shaft encoders. Figure 4.4 explains the wheel configuration for two such wheel configurations. The fixed standard wheel, steered standard wheel, or castor wheels cannot achieve omnidirectional motion.

There are two different wheel configurations to achieve omnidirectional movement.

1.   Swedish wheel
2.   Spherical wheel

Swedish wheels don't have a vertical axis of rotation. However, they can achieve omnidirectional motion like a castor wheel. This is possible by adding a degree of

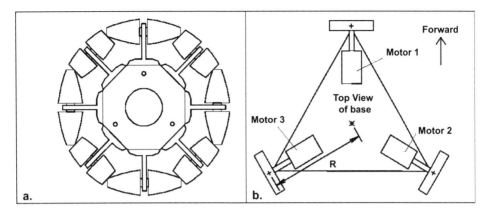

FIGURE 4.4   **a. Schematic of the wheel assembly used by the Veterans administration on an omnidirectional wheel chair. b. Top view of base showing relative orientation of components in the three-wheel configuration.**

freedom to the fixed standard wheel with rollers attached to the fixed standard wheel perimeter with axes that are antiparallel to the main axis of the fixed wheel component. The exact angle between the roller axis and the main axis can vary.

A spherical wheel does not put any constraint on the robot chassis' kinematics. Such a mechanism has no principal axis of rotation, and therefore no appropriate rolling or sliding constraints exist. The omnidirectional spherical wheel can have any arbitrary direction of movement. This kind of wheel imparts complete omnidirectional property to the robot chassis. However, this type of wheel is strictly applicable for indoor application only.

The omnidirectional wheel configuration shown in Figure 4.4(b) is based on three wheels, each actuated by one motor. The robot has three Swedish 90-degree wheels, arranged radially symmetrically, with the rollers perpendicular to the each main wheel.

4.2.4   Synchro Drive WMRs

An innovative configuration known as synchro drive features three or more wheels mechanically coupled in such a way that all rotate in the same direction at the same speed, and similarly pivot in unison about their respective steering axes when executing a turn. This drive and steering "synchronization" results in improved odometry accuracy through reduced slippage, since all wheels generate equal and parallel force vectors at all times.

The required mechanical synchronization can be accomplished in a number of ways; the most common being a chain, belt, or gear drive. Carnegie Mellon University has implemented an electronically synchronized version on one of their

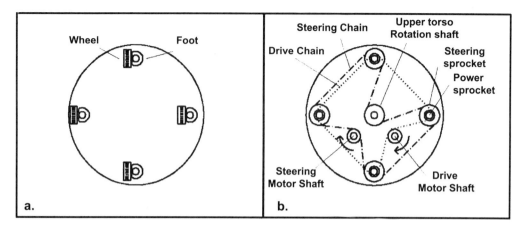

FIGURE 4.5   A four-wheel syncro-drive configuration: a. Bottom view, b. Top view.

Rover-series robots, with dedicated drive motors for each of the three wheels. Chain- and belt-drive configurations experience some degradation in steering accuracy and alignment due to uneven distribution of slack, which varies as a function of loading and direction of rotation. In addition, whenever chains (or timing belts) are tightened to reduce such slack, the individual wheels must be realigned. These problems are eliminated with a completely enclosed gear-drive approach. An enclosed gear train also significantly reduces noise as well as particulate generation, the latter being very important in clean-room applications.

Referring to Figure 4.5, drive torque is transferred down through the three steering columns to polyurethane-filled rubber tires. The drive-motor output shaft is mechanically coupled to each of the steering-column power shafts by a heavy-duty timing belt to ensure synchronous operation. A second timing belt transfers the rotational output of the steering motor to the three steering columns, allowing them to synchronously pivot throughout a full 360-degree range. The sentry's upper head assembly is mechanically coupled to the steering mechanism in a manner similar to that illustrated in Figure 4.5, and thus always points in the direction of forward travel. The three-point configuration ensures good stability and traction, while the actively driven large-diameter wheels provide more than adequate obstacle climbing capability for indoor scenarios. The disadvantages of this particular implementation include odometry errors introduced by compliance in the drive belts as well as by reactionary frictional forces exerted by the floor surface when turning in place.

To overcome these problems, the Cybermotion K2A Navmaster robot employs an enclosed gear-drive configuration with the wheels offset from the steering axis as shown in Figure 4.6. When a foot pivots during a turn, the attached

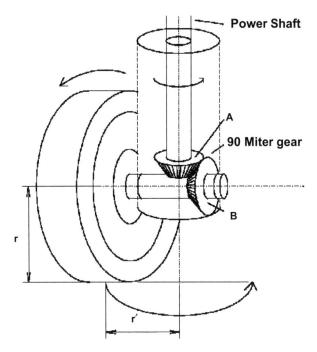

FIGURE 4.6   **Slip compensation during a turn is accomplished through use of an offset foot assembly on the three-wheeled K2A Navmaster robot.**

wheel rotates in the appropriate direction to minimize floor and tire wear, power consumption, and slippage. Note that for correct compensation, the miter gear on the wheel axis must be on the opposite side of the power shaft gear from the wheel as illustrated.

4.3   KINEMATICS AND MATHEMATICAL MODELING OF WMRS

4.3.1   What is Mathematical Modeling?

Mathematics provides a set of ideas and tools that are effective in solving problems, which arise in other fields. When used in problem solving, mathematics may be applied to specific problems already posed in mathematical form, or it may be used to formulate such problems. When used in theory construction, mathematics provides abstract structures that aid in understanding situations arising in other fields. Problem formulation and theory construction involve a process known as mathematical model building. Given a situation in a field other than mathematics or in everyday life, mathematical model building is the activity that begins with the situation and formulates a precise mathematical problem whose solution, or analysis in the case of theory construction, enables us to better understand the original situation.

Mathematical modeling usually begins with a situation in the real world, sometimes in the relatively controlled conditions of a laboratory and sometimes in the much less completely understood environment of meadows and forests, offices and factories, and everyday life. For example, a psychologist observes certain types of behavior in rats running in a maze, a wildlife ecologist notes the number of eggs laid by endangered sea turtles, or an economist records the volume of international trade under a specific tariff policy. Each seeks to understand the observations and to predict future behavior. This close study of the system, the accumulation and organization of information is really the first step in model building.

The next step is an attempt to make the problem as precise as possible. One important aspect of this step is to identify and select those concepts to be considered as basic in the study and to define them carefully. This step typically involves making certain idealizations and approximations. This step of identification, idealization, and approximation will be referred to as constructing a real model.

The third step is usually much less well defined and frequently involves a high degree of creativity. The goal is the expression of the entire situation in symbolic terms. As a consequence, the real model becomes a mathematical model in which the real quantities and processes are replaced by mathematical symbols and relations (sets, functions, equations, etc.) and mathematical operations. Usually, much of the value of the study hinges on this step because an inappropriate identification between the real world and a mathematical structure is unlikely to lead to useful results. It should be emphasized that there may be several mathematical models for the same real situation. In such circumstances, it may be the case that one model accounts especially well for certain observation while another model accounts for others.

After the problem has been transformed into symbolic terms, the resulting mathematical system is studied using appropriate mathematical ideas and techniques. The results of the mathematical study are theorems, from a mathematical point of view, and predictions, from the empirical point of view. The important contribution of the study may well be the recognition of the relationship between known mathematical results and the situation being studied.

The final step in the model-building process is the comparison of the results predicted on the basis of the mathematical work with the real world. The most desirable situation is that the phenomena actually observed are accounted

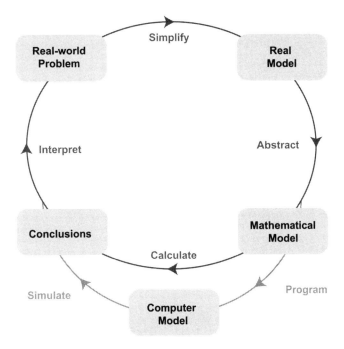

FIGURE 4.7

for in the conclusions of the mathematical study and that other predictions are subsequently verified by experiment. A typical situation would be that the set of conclusions of the mathematical theory contains some which seem to agree and some, which seem to disagree with the outcomes of experiments. In such a case one has to examine every step of the process again. It usually happens that the model-building process precedes through several iterations, each a refinement of the preceding, until finally an acceptable one is found. Pictorially, we can represent this process as in Figure 4.7.

The solid lines in the figure indicate the process of building, developing, and testing a mathematical model as we have outlined it above. The dashed line is used to indicate an abbreviated version of this process, which is often used in practice.

4.3.2 Kinematic Constraints

Figures 4.8 and 4.9 show how the instantaneous center of rotation is derived from the robot's pose (in the case of a car-like mobile robot) or wheel velocities (in the case of a differentially driven robot). The magnitude of the instantaneous rotation is in both cases determined by the magnitudes of the wheel speeds; the distance between the instantaneous center of rotation and the

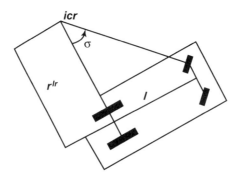

FIGURE 4.8   **Instantaneous center of rotation (icr) for a car-like robot.**

wheel center points is called the steer radius, [18], or instantaneous rotation radius. Figures 4.8 and 4.9 and some simple trigonometry show that

$$r^{ir} = \begin{cases} \dfrac{1}{\tan(\sigma)}, & \text{for a car-like robot,} \\[2ex] \dfrac{d}{2}\dfrac{v_r + v_l}{v_r - v_l}, & \text{for a differentially driven robot} \end{cases} \qquad (4.1)$$

with the wheelbase of the car-like robot, [18], (i.e., the distance between the points where both wheels contact the ground), the steer angle, the distance between the wheels of the differentially driven robot, and its wheel velocities.

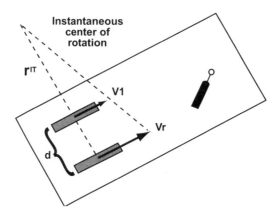

FIGURE 4.9   **Instantaneous center of rotation for a differentially driven robot.**

Differentially driven robots have two instantaneous degrees of motion freedom, compared to one for car-like robots. A car-like mobile robot must drive forward or backward if it wants to turn but a differentially driven robot can turn on the spot by giving opposite speeds to both wheels. In practice, the instantaneous rotation center of differentially driven robots can be calculated more accurately than that of car-like robots, due to the absence of two steered wheels with deformable suspensions.

4.3.3   Holonomic Constraints

Let us consider a robot A having an m-dimensional configuration space C. Let us now suppose that at any time t, we impose an additional scalar constraint of the following form to the configurations of the robot:

$$F(q,t) = F(q_1,q_2...,q_m,t) = 0, \qquad (4.2)$$

where F is a smooth function with a nonzero derivative. This constraint selects a subset of configurations of C (those which satisfy the constraint) where the robot is allowed to be.

We can use the equation (1) to solve for one of the coordinates, say q_m, by expressing it as a function g of the m-1 remaining coordinates and time, i.e., $g(q_1,.., q_m -1,t)$. The function g is smooth so that the equation (1) defines an (m-1)-dimensional smooth submanifold of C. This submanifold is in fact the configuration space of A and the m-1 remaining parameters are the coordinates of the configuration q.

Definition: A scalar constraint of the form F(q,t)=0, where F is a smooth function with a nonzero derivative, is called a holonomic equality constraint.

More generally, there may be k holonomic equality constraints (k<=m). If they are independent (i.e., their Jacobian matrix has rank k) they determine an (m-k)-dimensional submanifold of C, which is the actual configuration space of A.

Typical holonomic constraints are those imposed by the prismatic and revolute joints of a manipulator arm.

4.3.4   Nonholonomic Constraints

If a system has restrictions in its velocity, but those restrictions do not cause restrictions in its positioning, the system is said to be nonholonomically constrained. Viewed another way, the system's local movement is restricted, but not its global movement. Mathematically, this means that the velocity constraints cannot be integrated to position constraints. The most familiar example of a nonholonomic system is demonstrated by a parallel parking maneuver. When a driver arrives next to a parking space, he cannot simply slide his car sideways into the spot. The car is not capable of sliding sideways and this is the velocity restriction. However, by moving the car forward and backward and turning the wheels, the car can be placed in the parking space. Ignoring the restrictions caused by external objects, the car can be located at any position with any orientation, despite lack of sideways movement.

Let us consider the robot A while it is moving. Its configuration q is a differentiable function of time t. We impose that A's motion satisfy a scalar constrain of the following form:

$$G(q,q',t) = G(q_1,\ldots, q_m, q_1',\ldots, q_m',t) = 0, \tag{4.3}$$

where G is a smooth function and $q_i' = dq_i/dt$ for every i = 1,..., m. The velocity vector $q' = (q_1',\ldots, q_m')$ is a vector of $T_q(C)$ the tangent space of C at q. In the absence of kinematic constraints of the form (2), the tangent space is the space of the velocities of A.

A kinematic constraint of the form (2) is holonomic if it is integrable, i.e., if all the velocity parameters q_1' through q_m' can be eliminated and the equation (2) rewritten in the form (1). Otherwise, the constraint is called a nonholonomic constraint.

Definition: A nonintegrable scalar constraint of the form:

$$G(q_1,\ldots q_m. q_1',\ldots, q_m',t) = 0,$$

where G is a smooth function, is called a nonholonomic equality constraint.

Example

Consider a car-like robot (i.e., a four-wheel front-wheel-drive vehicle) on a flat ground. We model this robot as a rectangle moving in $W=R^2$, as illustrated in Figure 4.10. Its configuration space is R^2*S^1. We represent a configuration as a triple

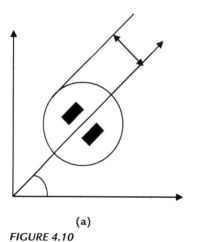
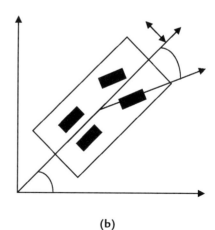

(a)   (b)

FIGURE 4.10

(x,y,θ) where (x,y)ε R^2 are the coordinates of the midpoint R between the two rear wheels and θε[0,2π] is the angle between the x-axis of the frame F_w attached to the workspace and the main axis of the car. We assume that the contact between each of the wheels and the ground is a pure rolling contact between two perfectly rigid bodies. When the robot moves the point R describes a curve γ that must be tangent to the main axis of the car. Hence, the robot's motion is constrained by:

$$-x' \sin(\theta) + y' \cos(\theta) = 0. \tag{4.4}$$

4.3.5 Equivalent Robot Models

Real-world implementations of car-like or differentially driven mobile robots have three or four wheels, because the robot needs at least three noncollinear support points in order to not fall over. However, the kinematics of the moving robots are most often described by simpler *equivalent* robot models: a "*bicycle*" robot for the car-like mobile robot (i.e., the two driven wheels are replaced by one wheel at the midpoint of their axle, whose velocity is the mean v_m of the velocities v_l and v_r of the two real wheels) and a "caster-less" robot for the differentially driven robot (the caster wheel has no kinematic function; its only

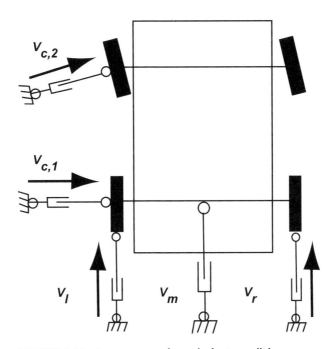

FIGURE 4.11 **Instantaneously equivalent parallel manipulator models for a car-like robot.**

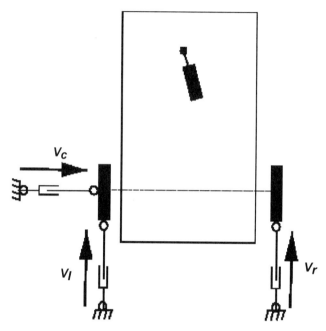

FIGURE 4.12 **Instantaneously equivalent parallel manipulator models for a differentially driven robot.**

purpose is to keep the robot in balance). In addition, Figures 4.11 and 4.12 show how an equivalent (planar) parallel robot can model car-like and differentially driven mobile robots. The nonholonomic constraint is represented by a zero actuated joint velocity v_c in the leg on the wheel axles. A car-like robot has two such constraints; a differentially driven robot has one. Since the constraint is nonholonomic and hence not integrable, the equivalent parallel robot is only an *instantaneous* model, i.e., the base of the robots moves together with the robots. Hence, the model is only useful for the velocity kinematics of the mobile robots. The velocities in the two kinematic chains on the rear wheels of the car-like robot are not independent; in the rest of this Chapter they are replaced by one single similar chain connected to the midpoint of the rear axle.

The equivalent car-like robot model is only an approximation, because neither of the two wheels has an orientation that corresponds exactly to the steering angle. In fact, in order to be perfectly outlined, a steering suspension should orient both wheels in such a way that their perpendiculars intersect the perpendicular of the rear axle at the same point. In practice, this is never perfectly achieved, so one hardly uses car-like mobile robots when accurate motion is desired. Moreover, the two wheels of a real car are driven through a *differential gear transmission*, in order to divide the torques over both wheels in such a way that neither of them slips. As a result, the mean velocity of both wheels is the velocity of the drive shaft.

In the following sections we will construct the kinematic models of the above two types of WMRs and develop appropriate control strategies for them.

4.3.6 Unicycle Kinematic Model

A differentially driven wheeled mobile robot is kinematically equivalent to a unicycle. The model discussed here is a unicycle-type model having two rear wheels driven independently and a front wheel on a castor. The following kinematic model is constructed with respect to the local coordinate frame attached to the robot chassis. The kinematic model for the nonholonomic constraint of pure rolling and nonslipping is given as follows:

$$q_d = S(q) * v. \tag{4.5}$$

Where q(t), qd(t) are defined as,

$$q = \begin{bmatrix} x_c \\ y_c \\ \theta_c \end{bmatrix} \tag{4.6}$$

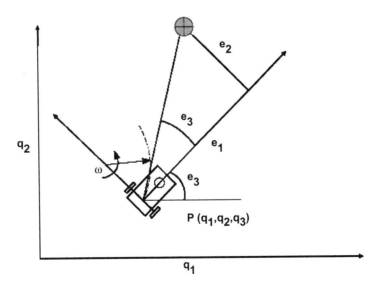

FIGURE 4.13   Relevant variables for the unicycle (top view).

$$\dot{q} = \begin{bmatrix} \dot{x}_c \\ \dot{y}_c \\ \dot{\theta}_c \end{bmatrix} \qquad (4.7)$$

$x_c(t)$ and $y_c(t)$ denote the position of the center of mass of the WMR along the X and Y Cartesian coordinate frames and $\theta(t)$ represents the orientation of the WMR, $x_{cd}(t)$ and $y_{cd}(t)$ denote the Cartesian components of the linear velocity, the matrix S(q) is defined as follows:

$$S(q) = \begin{bmatrix} \cos\theta & 0 \\ \sin\theta & 0 \\ 0 & 1 \end{bmatrix}. \qquad (4.8)$$

And velocity vector v(t) is defined as

$$v = \begin{bmatrix} v_1 \\ v_2 \end{bmatrix} = \begin{bmatrix} v_1 \\ \theta_d \end{bmatrix}. \qquad (4.9)$$

The control objective of regulation problem is to force the actual Cartesian position and orientation to a constant reference position and orientation. To quantify the regulation control objective, we define x(t), y(t), and θ(t) as the difference between the actual Cartesian position and orientation and the reference position as follows:

$$x(t) = x_c - x_{rc} \qquad (4.10)$$

$$y(t) = y_c - y_{rc} \qquad (4.11)$$

$$\theta(t) = \theta_c - \theta_{rc}. \qquad (4.12)$$

x_{rc}, y_{rc}, and θ_{rc} represent the constant position and orientation.

$$q_1 = x\cos\theta + y\sin\theta \qquad (4.13)$$

$$q_2 = -x\sin\theta + y\cos\theta \qquad (4.14)$$

$$q_3 = \theta \qquad (4.15)$$

Where q_1, q_2, and q_3 are the auxiliary error of the system. Taking the derivatives of the above and using the kinematic model given in equation (4.7), it can be rewritten as follows:

$$\dot{q}_1 = v_1 + v_2 e_2 \qquad (4.16)$$

$$\dot{q}_2 = -v_2 e_2 \tag{4.17}$$

$$\dot{q}_3 = v_2. \tag{4.18}$$

v_1 = The longitudinal velocity applied to the vehicle.
v_2 = The instantaneous angular velocity of the chassis of the vehicle.

The controls for this model are developed in Section 4.4.6.

4.3.7   Global Coordinate Kinematic Model of the Unicycle

In this section we will construct kinematic models for unicycle- and car-type WMRs with respect to the global reference frame. Given a global reference plane in which the instantaneous position and orientation of the model is given by (q(1), q(2), q(3)) with respect to the global reference system. The vehicle is to start at a position (x, y, θ) and has to reach a given point (x_d, y_d, θ d) with respect to the global reference plane. We will discuss how it does this in the control section of this chapter.

The longitudinal axis of the reference frame is attached to the vehicle and the lateral axis is perpendicular to the longitudinal axis. Since this reference frame's position changes continuously with respect to the global reference system, the instantaneous position of the origin of the reference frame attached to the vehicle is given by (q_1, q_2, q_3). The position of the point to be traced in the reference frame attached to the vehicle, with respect to the global coordinate system, is given by (e_1, e_2, e_3). Where,

e_1 = The instantaneous longitudinal coordinate of the desired point to be traced with respect to the reference system of the vehicle.

e_2 = The instantaneous lateral coordinate of the desired point to be traced with respect to the reference system of the vehicle.

e_3 = The instantaneous angular coordinate of the desired point to be traced with respect to the reference system of the vehicle.

The conversions of the local values of

$$e_1 = (x_d - q_1)* \cos q_3 + (y_d - q_2)* \sin q_3 \tag{4.19}$$

$$e_2 = -(x_d - q_1)* \sin q_3 + (y_d - q_2)* \sin q_3 \tag{4.20}$$

$$e_3 = \tan^{-1}\left(\frac{y_d - q_2}{x_d - q_2}\right) - \theta . \tag{4.21}$$

The kinematic model for the so-called kinematic wheel under the nonholonomic constraint of pure rolling and nonslipping is given as follows.

$$\dot{q}_1 = v_1 * \cos q_3 \tag{4.22}$$

$$\dot{q}_2 = v_1 * \sin q_3 \tag{4.23}$$

$$\dot{q}_3 = v_2 \tag{4.24}$$

v_1 = The longitudinal velocity applied to the vehicle.
v_2 = The instantaneous angular velocity of the chassis of the vehicle.

So these two variables have to be controlled by a control strategy, so that the vehicle reaches the desired point smoothly. The controls for this model are developed in Section 4.4.6.

4.3.8   Global Coordinate Kinematic Model of a Car-type WMR

In this section, we will discuss the kinematic model of a car-type WMR. The model is modeled with respect to the global reference frame. Given a global reference plane in which the instantaneous position and orientation of the model is given by (q(1), q(2), q(3)) with respect to the global reference system. The vehicle is to start at a position (x, y, θ) and has to reach a given point (xd, yd, θd) with respect to the global reference plane. We will discuss how it does this in the control section of this chapter.

The longitudinal axis of the reference frame is attached to the vehicle and the lateral axis is perpendicular to the longitudinal axis. The instantaneous position of the origin of the reference frame attached to the vehicle is given by (q_1, q_2, q_3). The position of the point to be traced in the reference frame attached to the vehicle, with respect to the global coordinate system, is given by (e_1, e_2, e_3).

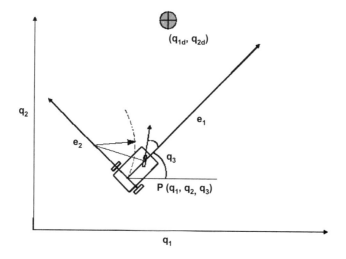

FIGURE 4.14   Relevant variables for the car-type model.

Where,

1. e_1 = The instantaneous longitudinal coordinate of the desired point to be traced with respect to the reference system of the vehicle.
2. e_2 = The instantaneous lateral coordinate of the desired point to be traced with respect to the reference system of the vehicle.
3. e_3 = The instantaneous angular coordinate of the desired point to be traced with respect to the reference system of the vehicle.

The conversions of the local values of:

$$e_1 = (x_d - q_1)* \cos q_3 + (y_d - q_2)* \sin q_3 \qquad (4.25)$$

$$e_2 = -(x_d - q_1)* \sin q_3 + (y_d - q_2)* \cos q_3 \qquad (4.26)$$

$$e_3 = \tan^{-1}\left(\frac{y_d - q_2}{x_d - q_2}\right) - \theta. \qquad (4.27)$$

The kinematic model for the so-called kinematic wheel under the nonholonomic constraint of pure rolling and nonslipping is given as follows:

$$\dot{q}_1 = v * \cos q_3 \qquad (4.28)$$

$$\dot{q}_2 = v * \sin q_3 \qquad (4.29)$$

$$\dot{q}_3 = \frac{v}{\rho} = \left(\frac{v}{l}\right) * \tan \delta. \qquad (4.30)$$

Here, $\rho = \dfrac{1}{\tan \delta}$, which is the instantaneous radius of curvature of the trajectory of the vehicle, and,

v = the longitudinal velocity applied to the vehicle.
δ = The instantaneous angular orientation provided to the steering wheel of the vehicle.

So these two variables have to be controlled by a control strategy, so that the vehicle reaches the desired point smoothly. The controls for this model are developed in Section 4.4.6.

4.3.9 Path Coordinate Model

The global model is useful for performing simulations and its use is described in Section 4.4.6. However, on the hardware implementation, the sensors cannot detect the car's location with respect to some global coordinates. The sensors can

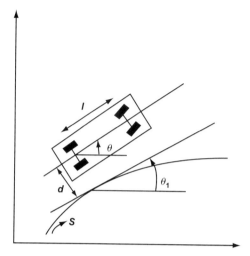

FIGURE 4.15   Path coordinate model for a car-type configuration.

only detect the car's location with respect to the desired path. Therefore, a more useful model is one that describes the car's behavior in terms of the path coordinates.

The path coordinates are shown in Figure 4.15. The perpendicular distance between the rear axle and the path is given by d. The angle between the car and the tangent to the path is $\theta_p = \theta - \theta_t$. The distance traveled along the path starting at some arbitrary initial position is given by s, the arc length.

The **kinematic model in terms of the kinematic model** is given by,

$$
\begin{bmatrix} \dot{s} \\ \dot{d} \\ \dot{\theta}_p \\ \dot{\phi} \end{bmatrix} = \begin{bmatrix} \dfrac{\cos \theta_p}{1 - dc(s)} \\ \sin \theta_p \\ \dfrac{\tan \phi}{l} - \dfrac{c(s)\cos \theta_p}{1 - dc(s)} \\ 0 \end{bmatrix} \upsilon_1 + \begin{bmatrix} 0 \\ 0 \\ 0 \\ 1 \end{bmatrix} \upsilon_2
$$

where c(s) is the path's curvature and is defined as,

$$
c(s) = \frac{d\theta_t}{ds}. \tag{4.31}
$$

We will not discuss the control law for this kinematic model in this text.

4.4   CONTROL OF WMRS

4.4.1   What is Control?

In recent years there has been great deal of research done in control of wheeled mobile robots. Most of these works have been concentrated on tracking and posture stabilization problems. The tracking problem is to design a control law, which makes the mobile robot follow a given trajectory. Stationary state feedback technique was used for this and many authors proposed stable controllers. The posture stabilization problem is to stabilize the vehicle to a given final posture starting from any initial posture (posture means both the position and orientation of a mobile robot from the base). The posture stabilization problem is more difficult than the tracking problem in the sense that nonholonomic systems with more degrees of freedom than control inputs cannot be stabilized by any static state feedback control law.

The basic motion tasks are redefined here for easy reference later.

Posture stabilization or point-to-point motion: The robot must reach a desired goal configuration starting from a given initial configuration.

Trajectory tracking: A reference point on the robot must follow a trajectory in the Cartesian space (i.e., a geometric path with an associated timing law) starting from a given initial configuration.

Execution of these tasks can be achieved using either open loop control, i.e., nonholonomic path planning or closed loop control, i.e., state feedback control, or a combination of the two. Indeed, feedback solutions exhibit an intrinsic degree of robustness. However, especially in the case of point-to-point motion, the design of closed loop control for nonholonomic systems has to face a serious structural obstruction. The design of open loop commands is instead strictly related to trajectory tracking, whose solution should take into account the specific nonholonomic nature of the WMR kinematics.

Open Loop Control

In nonholonomic path planning or open loop control, authors assume inputs (such as the linear speed and the steering angle or the angular motion of the

goal

start

FIGURE 4.16   **Basic motion tasks for a WMR: posture stabilization or point-to-point motion.**

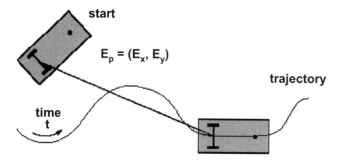

FIGURE 4.17   **Basic motion tasks for a WMR: trajectory tracking.**

WMR) in a kinematic model as the function of time, and then modify the assumed function to fit their purpose. Steering using sinusoidal input algorithms and iterative methods come under this category. For a differentially driven robot, this type of method is very easy to implement and is very efficient as well. But it is generally difficult to find or modify the inputs (linear speed and the steering angle), which transfer a car-like mobile robot to a desired posture, and hence a tracking control should be designed.

State Feedback Control

In a state feedback control system, the author assumes inputs (such as the linear speed and the steering angle or the angular motion of the WMR) in a kinematic model as the function of an error of the desired posture. Error is the difference between the instantaneous posture and the desired posture. As the WMR progresses toward its goal, it continuously modifies its course of motion until it finally reaches the desired posture. The most important merit of state feedback control in posture stabilization is that it can be directly used as a controller without any path planning. This type of control is more suitable for a car-type WMR than other types.

4.4.2   Trajectory Following

The objective of a kinematics controller is to follow a trajectory described by its position or velocity profiles as a function of time. This is often done by dividing the trajectory (path) in motion segments of a clearly defined shape, for example, straight line segments and segments of a circle. The control strategy is thus to precompute a smooth trajectory based on lines and circle segments, which drives the robot from the initial position to the final position. This approach can be regarded as open loop motion control, because the measured robot position is not feedback for velocity or position control. It has several disadvantages:

- It is not at all easy to precompute a feasible trajectory if all limitations and constraints of the robot's velocities are to be considered.
- The robot will not automatically adapt or correct the trajectory if dynamic changes in the environment occur.
- Resulting trajectories are not usually smooth, because the transmission from one trajectory segment to another are, for most of the commonly used segments (e.g., lines and parts of circles), not smooth. This means there is a discontinuity in the robot's acceleration.

We will discuss the control of a car-type mobile robot using the trajectory following approach here.

This can be summarized as follows. The space in the vehicle frame of reference is divided into a number of different geometric regions. The behavior of the vehicle can be modeled in a particular way according to the presence of the point in a particular region of space until it reaches a very close vicinity of the desired point when it finally stops. Thus, the behavior of the vehicle in this model is pretty predictable and hence various control strategies can be applied to the model easily. The model can be described as follows.

The Model

The model as shown in Figure 4.18 is very geometric. The space is divided into five discrete regions with respect to the vehicle frame of reference. These regions can be defined as follows.

1. This portion can be defined in the vehicle Cartesian coordinate reference plane as,

$$x_l \geq 0, \text{ and} - \varepsilon \leq y_l \leq \varepsilon.$$

   This is the thin rectangular strip of width 2ε along the longitudinal axis in the vehicle frame of reference.

2. This portion can be defined in the vehicle Cartesian coordinate reference plane as,

$$y_l > \varepsilon, \text{ for } x_l \geq 0$$

$$y_l \geq 0, \text{ for } x_l \leq 0$$

$$\text{and, } x_l^2 + (y_l - r_b)^2 > r_b^2.$$

   This is the entire positive y plane, except for portion 1 and 3.

3. This portion can be defined in the vehicle Cartesian coordinate reference plane as,

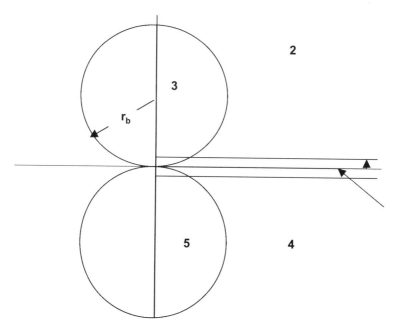

FIGURE 4.18 **The diagrammatic representation of the model and the different regions of space associated with it.**

$$x_l^2 + (y_l - r_b)^2 \leq r_b^2.$$

This is the portion inside a circle of radius r_b in the positive y plane, as shown in the figure.

4. This portion can be defined in the vehicle Cartesian coordinate reference plane as,

$$y_l > \varepsilon, \text{ for } x_l \geq 0$$

$$y_l \leq 0, \text{ for } x_l \leq 0$$

$$\text{and, } x_l^2 + (y_l - r_b)^2 > r_b^2.$$

This is the entire negative y plane, except for portion 1 and 3.

5. This portion can be defined in the vehicle Cartesian coordinate reference plane as,

$$x_l^2 + (y_l - r_b)^2 \leq r_b^2.$$

This is the portion inside a circle of radius r_b in the positive y plane, as shown in Figure 4.18.

4.4.3   The Control Strategy

The control objective is to design a controller for the kinematic model given in Section 4.3.7 that forces the actual Cartesian position and orientation to a constant reference position and orientation. Based on this control objective, a simple time-varying controller was proposed as follows.

$$\text{if } (e_1^{\,2} + e_2^{\,2} \geq c^2)^2$$
$$v = V_0$$

else

$$v = V_0$$

end.

This means that the velocity is a constant and has a value $v = V_0$, for all the points in the space except for the points inside a circle of radius c. This circle is the region in which we can choose the vehicle to finally stop. This can be chosen as small as required for the vehicle to stop at a very close vicinity of the desired point.

For the angular velocity control, the following conditional control strategy is adopted. The angular velocity fed to the system is region specific. A step angular displacement value is fed to the system as follows:

1.   For region 1 the angular deviation is,

$$\delta = 0;$$

2.   For regions 2 and 5 the value of the angular deviation is,

$$\delta = \delta_0;$$

3.   For regions 3 and 4 the value of the angular deviation is,

$$\delta = -\delta_0;$$

This control strategy is simulated using MATLAB. The simulation and test results are discussed in Section 4.5.

4.4.4   Feedback Control

A more appropriate approach in motion control of a wheeled mobile robot is to use a real-state feedback controller. With such a controller, the robot's path-planning task is reduced to setting intermediate positions (subgoals) lying on the requested path. We will discuss the control of a differentially driven mobile robot and a car-type mobile robot using the feedback control approach here.

Developing Control for a Unicycle-type Mobile Robot

In this section we will discuss a control strategy developed for a unicycle (differentially driven) kinematic model, and implement it on the kinematic model discussed in Section 4.3.6. The control objective is to design a controller for the transformed kinematic model given by equations 4.16, 4.17, and 4.18 that forces the actual Cartesian position and orientation to a constant reference position and orientation. Based on this control objective, a differentiable, time-varying controller was proposed as follows:

$$v_1 = -k_1 e_1 \tag{4.32}$$

$$v_2 = -k_2 e_3 + e_2{}^2 \sin t \tag{4.33}$$

where k_1 and k_2 are positive constant control gains. After substituting the equations the following closed-loop error system was obtained:

$$\dot{q}_1 = -k_1 q_1 + (-k_2 q_3 + q_2{}^2 \sin t) \cdot q_2 \tag{4.34}$$

$$\dot{q}_2 = -(-k_2 q_2 + q_2{}^2 \sin t) \cdot q_2 \tag{4.35}$$

$$\dot{q}_3 = (-k_2 q_3 + q_2{}^2 \sin t). \tag{4.36}$$

The control strategy adopted here is quiet simple. The linear velocity is directly proportional to the longitudinal error or the projected distance of the vehicle from the destination point alone. The rate at which the wheels should be turned is proportional to the angular orientation of the desired point with respect to the reference frame attached to the vehicle or the angular error and an additional time-varying term. This term plays a key role when the vehicle gets stuck at a point. Such a situation occurs when the longitudinal error of the vehicle vanishes and it is oriented parallel to the desired direction, but the lateral error is not zero. In such a case, in the absence of the second term of the angular velocity control, the vehicle would get locked in that position and will fail to move any further even though the vehicle has not reached the destination point. So this additional term is added to the steering control term. When the vehicle gets locked in the above-mentioned situation, this term makes the angular error nonzero again and makes the vehicle turn a bit. This gives rise to a longitudinal error and the velocity is again nonzero. The term is a sine function of time multiplied with the square of the lateral error. The sine term varies between −1 and 1 making this term vary in an oscillatory fashion. The lateral error term makes the quantity bigger when the lateral error is large. So, when the lateral error is large the disturbing steer is even larger. This quantity is smaller in comparison to the first quantity, so that the cyclic nature of the sine function does not affect the result much. Here k_1 and k_2 are positive constant control gains. The above model will be simulated using Matlab in Section 4.5.

Developing Control for a Car-type Mobile Robot

In this section we will discuss a control strategy developed for the kinematic model of a car-type mobile robot discussed in Section 4.3.8. The control objective is to design a controller for the above kinematic model that forces the actual Cartesian position and orientation to a constant reference position. It is important to note that the control strategy discussed here is a basic control strategy and does not stabilize the vehicle to a desired orientation. It is therefore not very useful to implement in most real applications. Nevertheless, it is an easy control strategy to start learning. Based on this control objective a simple time-varying controller was proposed as follows.

$$v = vpar^* \, e_1 \tag{4.37}$$

$$v = vpar^* \, e1 \tag{4.38}$$

This simply means that, the longitudinal velocity is directly proportional to the longitudinal error in the reference system attached to the vehicle. The rate at which the wheels should be turned is proportional to the angular orientation of the desired position with respect to the reference frame attached to the vehicle. Here vpar and cpar are positive constant control gains. After substituting equations 4.37 and 4.38 into equations 4.28, 4.29, and 4.30, the following closed loop error system was developed.

$$\dot{q}_1 = vpar^* \, e_1 * \cos q_3 \tag{4.39}$$

$$\dot{q}_2 = vpar^* \, e_1 * \sin q_3 \tag{4.40}$$

$$\dot{q}_3 = \left(\frac{vpar * e_1}{l} \right) * \tan(cpar * e_3) \tag{4.41}$$

The above model will be simulated using MATLAB in Section 4.5.

4.5   SIMULATION OF WMRS USING MATLAB

The behavior of the model and the control strategy can be tested if we can obtain the trajectory of the path, when subjected to a given set of conditions. For that we need to get all the values of the state variables (q_1, q_2, q_3) at small intervals of time, which can later be plotted to obtain the trajectory of the path followed. Hence the above set of first-order differential equations have to be integrated in a time interval; given the values of the initial conditions and parameters using ode23—a powerful tool of Matlab. Ode23 is a function for the numerical solution of ordinary differential equations. It can solve simple differential equations or simulate complex dynamical systems. It integrates a system of ordinary differential equations using second- and third-order Runge-Kutta formulas. This particular third-order method reduces to Simpson's 1/3 rule and uses the third-order estimate for xout. The process of ode23 is as follows: A string variable with the name of the M-file that defines the differential equations to be integrated. The function needs to compute the state derivative vector, given the current time and state vector. It must take two input arguments; scalar t (time) and column vector q (state), and return output argument qdot, a column vector of state derivatives. The above set of first-order differential equations was converted into the following M-file, to execute ode23.

4.5.1   Testing the Control Strategy for a Unicycle-type Mobile Robot

The control strategy developed in this section is modeled in MATLAB and is tested for the vehicle to reach different destination points. In this example, we will discuss in detail how the simulation results can guide the researcher to predict a modification in the control strategy so as to optimize the result. The following plot is a typical result of simulation, which shows the trajectories of the state variables q_1 and q_2. The plot presented here is not a visualization of the actual model. The trace is plotted with respect to the local reference frame attached

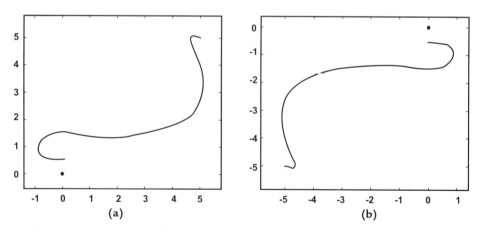

FIGURE 4.19   (a) Vehicle starting from (5, 5, 0); $k_1 = 1$, $k_2 = 1$; (b) Vehicle starting from (−5, −5, 0).

to the vehicle chassis. So the destination points in the plots are situated at zero and the starting points are the longitudinal and lateral errors. In the following plots in Figure 4.19, the vehicle starts with error values e_1 and e_2 as, (5, 5). The parameters are taken as, $k_1 = 1$ and $k_2 = 1$.

The Optimal Values of the Parameters and k_2

The above result shows the plots of the lateral against the longitudinal errors of the vehicle when the parameters are chosen to be $k_1 = 1$ and $k_2 = 1$. The values of k_1 and k_2 can be iterated to study their influence over the results and find out their optimal values that give the best results. From the iteration, the best values of the parameters were found to be $k_1 = 2$ and $k_2 = 0.1$. A comparison between the two results is shown in Figure 4.20.

From the figures it is clear that the lateral error (e_2) could not be completely eliminated even for the optimal values of the parameters k_1 and k_2. Still there remained a lateral error of about 0.5 units and it did not seem to improve much by the change of the values of the parameters. The amplitude of the longitudinal oscillations is, however, minimized; but the primary objective was to reduce the lateral error, which did not seem to be much affected by a variation in the parameters. The trajectory has, however, become reasonably smoother in the plot than the previous result. The vehicle now undergoes lesser wandering before reaching close to the goal position (see Figure 4.20). The reason for this lesser wandering of the vehicle accounts for the following: By choosing a smaller value of k_2 with respect to k_1, we are assigning lower weight to the angular error and giving more weight to the linear errors. This directly reduces the steering rate of the vehicle. Hence, the vehicle remains in more control. To understand this, let us consider driving a real car. If the driver swings the steering wheel too fast in response to curves in the road or if the steering wheel is too free, the vehicle

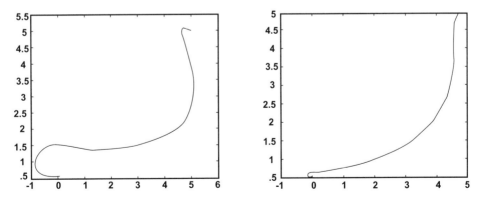

FIGURE 4.20   (a) The trajectory of the vehicle when, $k_1 = 1$; $k_2 = 1$; (b) The trajectory of the vehicle with the optimal values of the parameters, $k_1 = 2$; $k_2 = 0.1$.

experiences a lot of swagger in its motion. The motion is in control or smooth if he swings the wheel slowly, keeping a firm control over the steering wheel. This exactly happens here by choosing a lower value of the parameter k_2.

Plotting the trajectory for 100 seconds we find that the longitudinal error swings about 0 positions and the amplitude decreases continuously while the lateral error approaches 0 at a very slow rate. This is clear from Figure 4.21.

The enlarged view of the final oscillatory character of the trajectory suggests that the longitudinal error swings about 0, while the lateral error very slowly nears 0. The reason for this is explained as follows. The angular velocity is the sum of two components; $-k_2 e_3$ and $e_2^2 \sin t$. The presence of the sine term explains the orderly oscillations at the end part of the motion. Since the longitudinal and lateral errors become insignificant as the vehicle slowly proceeds toward the goal position, the sine term becomes the predominant factor in effecting the motion of the vehicle. This makes angular speed vary regularly with time. Since the longitudinal speed decreases as the longitudinal error decreases (the vehicle approaches close to the goal position), the vehicle travels through lesser distance in the same time in which the sine term changes sign; hence resulting in decrease of the amplitude of the oscillations as the vehicle approaches the destination point.

A Simple Modification in the Control Strategy

So finally we tried to find out if any modification in the model could affect the results. A lot of modifications were tested and finally an introduction of a third

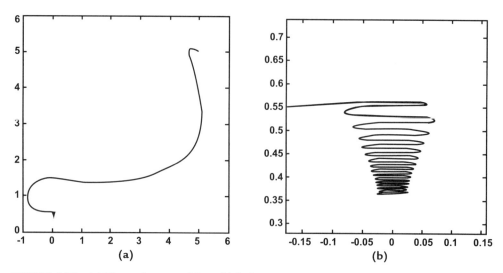

FIGURE 4.21   (a) The trajectory of the vehicle for 100 seconds: $k_1 = 1$, $k_2 = 1$; (b) The magnified view of the oscillatory nature of the motion.

constant, k_3, yielded desirable results. When a constant k_3 was multiplied with the sine terms, the lateral error seemed to almost reduce to zero at a much faster rate. The result was smoother trajectory and faster motion of the vehicle. Higher values of constant k_3 yielded even better results. The final oscillations seemed to almost come down to zero.

$$v_1 = -k_1 e_1 \tag{4.42}$$

$$v_2 = -k_2 e_3 + k_3 e_2^2 \sin t \tag{4.43}$$

The Results of the Modification

The modified strategy was tested for various values of state variables and parameters. The value of the parameter k_3 was iterated and it was observed that the final result was greatly improved with the increase in the value of k_3. For value of $k_3 = 10$, the final lateral error was reduced to a very small value and also at a much faster rate. This is shown in Figure 4.20(b) where, $k_1 = 2$ and $k_2 = 0.1$. The vehicle starts with error values (5, 5, 1) and is simulated for 10 seconds. Figure 4.22(b) shows the result for $k_3 = 100$ for the same values of k_1 and k_2.

It is clearly seen that the result is greatly improved by the application of this strategy. The result is also attained at a very fast rate. With the increase in the value of k_3, the result is also further improved. So in the following result the value of k_3 is taken to be 100.

From Figure 4.23, we can clearly see the final position of the trajectory. The lateral error is 0 and the longitudinal error is less than 0.005×10^{-3}, which is small enough to neglect. So it is clearly seen how the model got modified. The lateral error is 0 in less than 3 seconds. The speed of the vehicle at this point is also small enough to assume it to be 0.

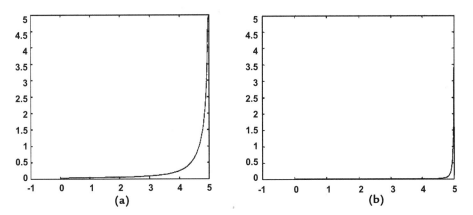

FIGURE 4.22   The trajectory of the vehicle with the modified strategy: (a) k_3 = 10; and (b) k_3 = 100.

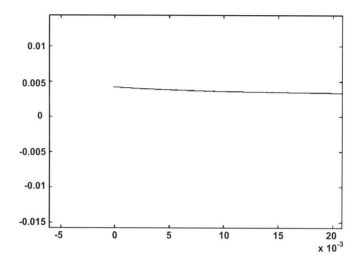

FIGURE 4.23   **The enlarged view.**

From the above discussion, it is clearly seen that the final result has been greatly improved by the modification in the strategy. By choosing a higher value of k_3, we are actually increasing the weight of the second term in the angular velocity control term. That means the angle now changes faster than before, and hence the vehicle reaches to the point at a faster rate. When the desired distance becomes very small, the angular change also becomes small. But by choosing higher values of k_3, it has become very fast and the amplitude has also increased, which makes the steps at which the vehicle nears the destination point bigger. But since the second term is a sine function of time it will simply fluctuate about a mean value, with amplitude diminishing slowly.

4.5.2   Testing the Control Strategy for a Car-type Mobile Robot

The control strategy developed in this section is modeled in MATLAB and is tested for the vehicle to reach different destination points. The above model and strategy was tested in MATLAB for various values of state variables and parameters. The MATLAB program is in Appendix II (a). Two plots are shown below in which the vehicle starts from (0, 0, 0) and reaches a destination point. The parameters: cpar = 1 and vpar = 1. The model is simulated for 10 seconds. The plots show the trajectory of the vehicle in a plane.

The Problems Encountered

The simulation and testing of the above model in MATLAB highlighted the following problems.

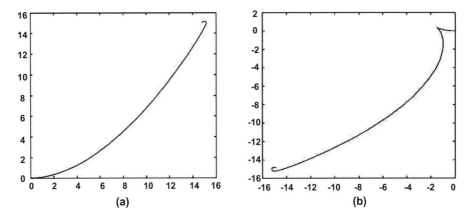

FIGURE 4.24   (a) Trajectory while tracing (5, 5); (b) Trajectory while tracing (-5, -5).

▓   The vehicle fails to start at all when the destination point lies on the y-axis i.e., when xdes = 0. That is because the value of ddist is zero when xdes is equal to zero. So, the velocity v, which is directly proportional to ddist, is also zero. Hence the vehicle does not start.

▓   After reaching sufficiently close to the destination point, the vehicle goes round about the point, instead of stopping.

These points are discussed below and the attempts taken toward solving those problems. Here the problems are shown diagrammatically.

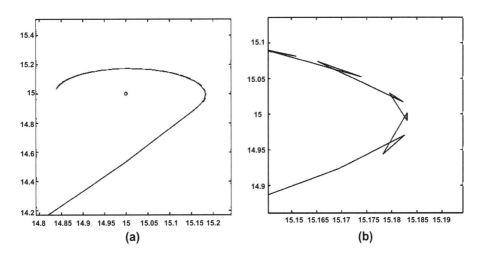

FIGURE 4.25   The above figure shows how the vehicle goes round the destination point.

In Figure 4.25, the above-mentioned problem of circling about the destination point is illustrated. The reason for this circling round the destination point is the following: The vehicle follows such a path that the longitudinal and lateral errors do not decrease proportionately. The longitudinal error decreases faster than the lateral error. So when the longitudinal error approaches zero, the lateral error becomes quiet large in comparison to it and simultaneously the angular error also becomes quiet large. The linear velocity of the vehicle varies proportionately with the longitudinal error, whereas the steering angular velocity is proportional to the linear velocity as well as the tangent of the angular error. Since the value of the tangent becomes quite large when the angle approaches $\pi/2$, the angular velocity of steering also increases indefinitely even though the linear velocity is very small. That means, even though the velocity decreases considerably when the longitudinal error approaches zero, the angular velocity value increases and makes the vehicle circle about the desired point. So the problem can be solved if the linear velocity and the steering angular velocity can be controlled in such a way that the vehicle goes straight to the destination point. That means, both the longitudinal and lateral error reduce proportionately. The control may be done in the following ways.

- By choosing suitable functions for vpar and cpar and thereby controlling the linear velocity and angular velocity in the desired way.
- By making certain improvements in the control strategy to deal with the problem, if satisfactory results were not obtained by the above two strategies.

The above strategies are applied and tested in the above model. The objective is to smoothen the motion of the vehicle and to remove all the above-mentioned problems so that finally, the vehicle should be able to reach all the points in the plane smoothly.

Modifications in the Control Strategy

A suitable value for vpar and cpar was found out by a lot of iterations. It was found that a value of vpar = 2 and cpar = 1.2 gives good results. This made the trajectory much smoother while reaching points sufficiently close to the origin.

The following modification is made here in the parent strategy to make the vehicle trace the points that lie along the lateral axis in the vehicle frame of reference. That is, by adding a small additional constant quantity to the previous equation of the linear velocity. So the linear velocity equation is modified to:

$$v = vpar^* \, e_1 + v{-}^0. \tag{4.44}$$

The angular velocity remains the same as in the previous strategy. This ensures that the vehicle would start at all the conditions. But then the problem now arisen due to this strategy is that the vehicle will never stop finally, even after reaching the point. It will continue to proceed in that direction with the same velocity v_0. This problem can be solved by choosing a strategy for the vehicle to finally stop once it comes within a given closeness of the destination point. The strategy is as in the following: the velocity of the vehicle greater than a particular error value would be according to the previous strategy and velocity of the vehicle for error less than that value is zero. The error value is the absolute distance of the instantaneous position value from the destination point. The strategy can be represented as follows:

$$\text{if (abs(dist) >= near)} \qquad \text{else}$$
$$v = vpar * e_1 + v\text{--}; \qquad v\text{-} = 0$$

where, dist = the absolute distance of the destination point from the instantaneous position of the vehicle and near = the required closeness at which the vehicle should stop.

The Result of Modifications

The modified strategy was developed in Matlab and tested for various values of state variables and parameters. Some plots are shown below in which the vehicle starts from (0, 0, 0) and reaches a point in the lateral axis in the vehicle frame of reference in both directions. That is, it is to reach (0, y) and (0, -y), where y is a variable. The parameters cpar and vpar are taken as cpar = 1 and vpar = 1. The parameter v_0 is iterated to find the most suitable value for it, so that the previous problem of getting stuck at certain points is solved satisfactorily. The strategy is

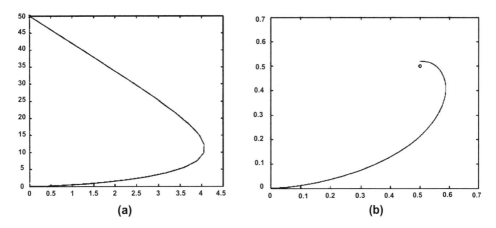

FIGURE 4.26   (a) Point to be traced is (0, 50): cpar = 1, and vpar = 1; (b) Point to be traced is (0.5, 0.5): cpar = 1, and vpar = 1.

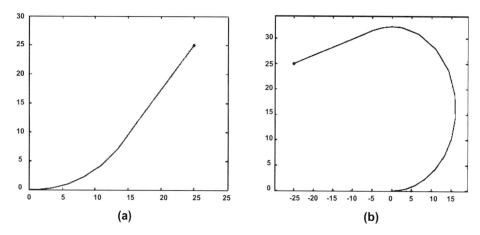

FIGURE 4.27   (a) Trajectory while tracing (25, 25); (b) Trajectory while tracing (-25, 25).

simulated for 10 seconds. Some plots showing some typical results are also shown for clearer visualization of the solution. The following plots show the trajectory of the vehicle in a plane. The points to be traced are marked with circles.

4.5.3   Testing the Control Strategy Trajectory Following Problem in a Car-type Mobile Robot

The control strategy developed in this section is modeled in MATLAB and is tested for the vehicle to reach different destination points. Two plots are shown in Figure 4.28 in which the vehicle starts from (0, 0, 0) and reaches a destination point. The parameters are: v0 = 10, err = 0.01, and c = 0.1. The model is simulated for 10 seconds. The plot shows the trajectory of the vehicle in a plane. The desired point is marked with an*.

The Problems Encountered

The model seems to perform quiet nicely from the results except for only one problem. The vehicle failed to finally stop at the destination point when the error range (error range is the minimum nearness to the destination point that the vehicle is finally required to attain) was smaller than a particular value, and that value was found to be dependent upon the strip width 'ε' of region 1. In the strategy the vehicle was required to go about a circle of radius 'r_b' until the desired point comes into region 1 in the vehicle frame of reference. Given the strategy lies in any region initially, the vehicle reaches region 1 quiet nicely. But as soon as it reaches very close to the desired point, instead of stopping at the required error range, the vehicle keeps on tracing circles indefinitely. The problem is shown

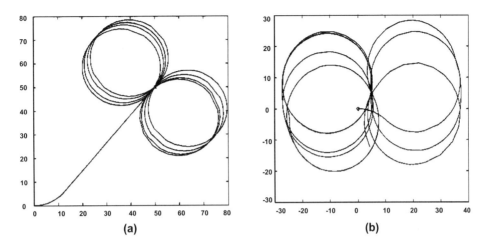

(a)   **(b)**

FIGURE 4.28   (a) The problem occurred when c was chosen to be 0.01, which is equal to ε; (b) The problem occurred when c was chosen to be 0.01, which is equal to ε, while tracing the point (5, 5). The starting point and the desired point are marked.

in the plots in Figure 4.28. In the plot for the strip width and the radius of the error-range, both are taken to be equal to 0.01.

4.6   The Identification and Elimination of the Problem

The problems were investigated for the possible reasons of failure. A number of predictions were made and the solutions were proposed. The first task was to isolate the region that contained the problem.

From a lot of iterative investigation of the above plots, it was observed that the region, which is the intersection of regions 1, 3, and 5, is the region where the problem occurs. To ensure this intuition the above-mentioned region was also included inside the error range, where the vehicle has to stop finally. Thus, from simple geometrical inspection, the modified error range was found to be,

$$c = \sqrt{2 \times r_b \times \varepsilon} \; . \tag{4.45}$$

This improvement in the strategy was implemented and simulated. The result clearly showed the problems removed. So the strategy changes a bit. Now in the modified strategy, the error range is not a parameter, but is predefined.

So the results clearly indicate that the problem is centered to the above-mentioned region only. After isolating the region where the problem was centered, the next task was to investigate the real reason why that problem occurs.

Finally, the problem was found to be the following: when the vehicle touches region 1, the angular velocity instantly becomes zero. So the desired point, instead of getting into region 1, lies at the boundary and slowly proceeds longitudinally. In the vehicle frame of reference, the desired point moves along the boundary toward the vehicle slowly. This happens because of the discontinuous nature of the angular velocity shift. So finally, when the desired point comes to the point where the border of the circular region intersects region 1, it behaves according to the conditions defined for being inside region 3. Hence, it attains an angular velocity of either $+\delta_0$ or $-\delta_0$ and instead of stopping there, it moves away from the destination point. This way it keeps on tracing circular trajectories indefinitely instead of reaching the desired point and stopping there. The problem can be solved if the point lies within region 1 instead of lying at the boundary. This can be done by another approach: by making the change of angular velocity continuous instead of discrete.

4.7   Modifying the Model to Make the Variation in Delta Continuous

In the previous strategy, the variation in angular changes was discrete, which is not possible practically. So the model needed a modification so as to make the process continuous. That means the angular variation will not take place in steps, but it will take place continuously. For this a fourth state variable q_4 is included in the model. This variable is the actual angular change that takes place when a step change in the desired angle change takes place. That means there is a time lag now between the required value of angular change and the actual change. This time-lag factor is taken care of by a parameter k. So now the angular change that takes place is, rather than. The modified model is given below.

$$\dot{q}_1 = v * \cos q_3 \tag{4.46}$$

$$\dot{q}_2 = v * \sin q_3 \tag{4.47}$$

$$\dot{q}_3 = \frac{v}{\rho} = \left(\frac{v}{l}\right) * \tan q_4 \tag{4.48}$$

$$\dot{q}_4 = \frac{(\delta_d - q_4)}{k} \tag{4.49}$$

Here, $\rho = \dfrac{l}{\tan \delta}$, which is the instantaneous radius of curvature of the trajectory of the vehicle, and

v = the longitudinal velocity applied to the vehicle.

δ_d = the desired value of instantaneous angular deflection provided to the wheels of the vehicle.

This δ_d is a function of the region in which it lies. It is the same step function that was for δ. It is defined as follows.

1. For region 1 the angular deviation is,

$$\delta = 0.$$

2. For regions 2 and 5 the value of the angular deviation is,

$$\delta_d = \delta_0.$$

3. For regions 3 and 4 the value of the angular deviation is,

$$\delta_d = -\delta_0.$$

The Result of the Simulations

The result of the simulations is shown in Figure 4.29. The vehicle starts from $(0, 0)$, and reaches the point $(45, 55)$. For values of err $= 0.01$ and c $= 0.01$. The same values showed problems previously. Now the problem seems to have been solved.

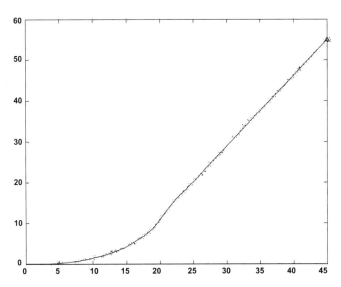

FIGURE 4.29 **A result showing the trajectory of the vehicle, finally successful for the desired values of parameters.**

4.8   Developing the Software and Hardware Model of an All-purpose Research WMR

In the following section, we will discuss the development of Turtle, a two-wheeled differentially driven wheeled mobile robot, which achieves various patterns of motion, by the differential combination of motion between both the wheels. The system is held in a statically stable posture with two driving wheels and two caster wheels. The hardware of the system consists of the following.

1. Two stepper motors.
2. Two gear reduction arrangements attached with both the wheels.
3. Two wheels attached with the gear reduction arrangements.
4. Two caster wheels.
5. A stepper motor controller circuit.

The objective of the project is to achieve the following.

1. Interfacing the above system with a parallel port.
2. Developing software to control the system using a parallel port and generate the following patterns of motion.

■   Motion in a straight line.
■   Motion along a circle.
■   Motion along a curve of any given radius of curvature.

Interfacing the System with a Parallel Port

Interfacing the above system with a parallel port requires the anatomy of the system. So first the system has to be analyzed properly. The existing system can be divided into the following three subsystems.

1. Mechanical
2. Electronic
3. Software

Mechanical Subsystem

The mechanical subsystem comprises of a 4-wheeled vehicle consisting of two driving wheels powered by two stepper motors and two castor wheels to support the vehicle. It also includes the structure, which houses the motors, batteries, the driver circuit, and reduction gearing mechanism.

The mechanical subsystem derives its input from the electronic subsystem, which is basically a stepper motor driving circuit. A schematic for this driving circuit is shown in Figure 4.30.

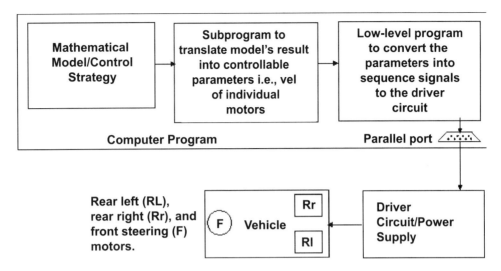

FIGURE 4.30   The relationship between various subsystems.

The software subsystem comprises of the mathematical model of the WMR, the various control strategies, the stepper motor driving algorithm, and the plotting functions used to represent the vehicle motion on the computer display.

Figure 4.30 shows the interrelationship between the three subsystems.

Electronic Subsystem

The electronic subsystem consists of a driver circuit to drive two unipolar stepper motors. A brief idea is given below. Each of the stepper motors requires four bits of data to drive it. These data energize the various motor coils in a particular sequence of patterns. Each pattern causes the motor to move one step. Smooth motion results from presenting the patterns in the proper order. A circuit or program, which is responsible for converting step and direction signals into winding energizing patterns, is called a *translator*. In our system the translator consists of the hardware of a computer and the C++ program to generate signals at the parallel port. Our stepper motor control system includes a *driver* in addition to the translator to handle the current drawn by the motor's windings.

Figure 4.31 shows a schematic representation of the "translator + driver" configuration. There are separate voltages for logic and for the stepper motor. The motor will require a different voltage than the logic portion of the system. Typically, logic voltage is +5 Vdc and the stepper motor voltage can range from +5 Vdc up to about +48 Vdc. The driver is also an "open collector" driver, wherein it takes its outputs to GND to activate the motor's windings.

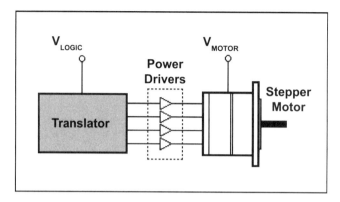

FIGURE 4.31 A typical translator/driver connection.

Interfacing with Parallel Port Using ULN2003 IC

Figure 4.32 presents a detailed schematic representation of the interfacing of the stepper motor with the parallel port using ULN2003 IC. Please refer back to Chapter 3 to revise the details about parallel port connections and the internal working of a ULN2003 IC.

The Driver Circuit

Another approach to control the stepper motor is by using a driver circuit. The driver interface to control the motor from is just a transistor switch replicated

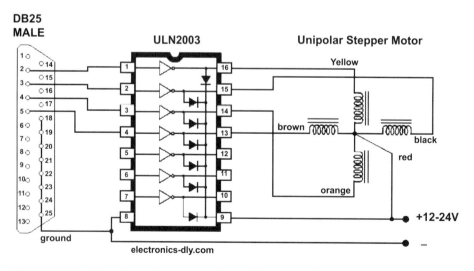

FIGURE 4.32 The schematic representation of the connections.

four times. The transistor controls the current, which is much higher than the parallel port sink capacity. This is done to allow for the motor voltage to be independent of the PC power supply. Figure 4.32 shows the schematic representation of the driver circuit that is used in our system. A positive voltage at the transistor base (writing a '1' to the appropriate bit at #Data) causes the transistor to conduct. This has the effect of completing the circuit by hooking up ground to the motor coil (which has a positive voltage on the other side). So the chosen coil is turned on.

Switching is one of the primary uses of transistors. A power transistor is used in the driver circuit so that it can switch lots of current (up to five amps for the TIP 120). A Darlington transistor is really a transistor pair in a single package with one transistor driving the other. A control signal on the base is amplified and then drives the second transistor. The resulting circuit cannot only switch large currents, but it can do so with a very small controlling current. The resistors are to provide current limiting through the parallel port. The diodes are a feature typical of circuits that handle magnetic coils that are *inductive* circuits. In this context, the motor windings are the inductive element. Inductors provide a means for storage of electrical *current*. The driving current causes a magnetic field to be built up in the coil. As soon as the drive is removed, the magnetic field collapses and causes the inductor to release its stored current. Semiconductors are particularly sensitive to these currents (they briefly become conductors and then become permanent nonconductors). The diodes provide a mechanism to safely shunt these currents away and, thus, protect the transistors and the computer.

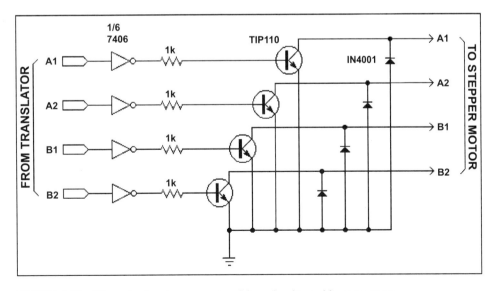

FIGURE 4.33   **The unipolar stepper motor driver circuit used in our system.**

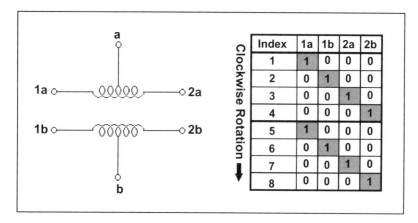

FIGURE 4.34　**Unipolar stepper motor coil setup (left) and 1-phase drive pattern (right).**

The Sequencing of the Stepper Motors

The stepper motor coils are required to be energized in a particular sequence. There are several kinds of sequences that can be used to drive stepper motors. The following table gives the sequence for energizing the coils that is used in the software of our system. The steps are repeated when reaching the end of the table. Following the steps in ascending order drives the motor in one direction; going in descending order drives the motor the other way.

The Software Subsystem

The software subsystem generates the signals required to drive the two stepper motors so that the vehicle is able to travel in the desired manner. This is attained in the following steps.

1. The user provides the desired destination points that the vehicle has to reach. The software designates an initial position to the vehicle and defines a final position that the vehicle has to reach in a Cartesian coordinate reference frame. Based on these values the software calculates a desired steering angle that the vehicle has to rotate and the desired distance that the vehicle has to travel.
2. Based on these values, the kinematic model of the system decides what wheel speeds have to be provided to the individual wheels. The kinematic model will be described in detail in the next section.
3. Finally, the stepper motor driving algorithm decides the stepping rate for the individual wheels.
4. The software interface generates a plot of the vehicle while in motion.

The Kinematic Model

The purpose of the kinematic model of the vehicle is to determine the relationship between the motions of the driving members of the system so that the motion is slip free. For a WMR, when the wheels do not skid, the motion is determined by the constraints of the geometry of the system. This kind of dynamic system is called a nonholonomic system. The mathematical model of a WMR gives the values of the actual vehicle speeds at the various wheels (the two rear wheels), when the vehicle is following a certain pattern of motion. These values are then implemented in the WMR to control its motion. The turtle is originally designed to be a differentially driven vehicle, which is the conventional design used in maximum robotic applications. However, the kinematics can be designed in an appropriate manner for the same hardware to behave like a car-type mobile robot also. A detailed description of both types of kinematic models will be given in the following sections.

Differentially Driven Wheeled Mobile Robot

The vehicle motion can be divided into three different modes: straight mode, steering mode, and combined motion mode. Any journey of the vehicle is actually composed of a number of straight and steering modes of travel. Considering a vehicle of length 'l' and width '2b,' the following relations can be established among the control parameters.

In the straight mode the vehicle travels in a straight line without any steering. In this case both wheels assume the same velocity.

$$vr = vl$$

In steering mode the vehicle steers either toward the left or right direction. The two driving wheels have to be provided motion in the desired manner following the constraints of motion. In the simplest following case, they are

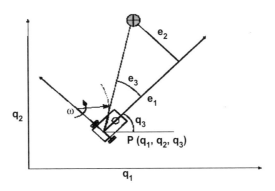

FIGURE 4.35   **The schematic representation of a differentially driven WMR.**

simply driven in opposite directions so that the vehicle revolves about its own center.

$$vr = -vl$$

or,

$$vl = -vr$$

In combined motion mode, the trajectory that the vehicle follows is an arc. In this mode, the vehicle progresses and simultaneously changes the direction of motion. The relative pattern of motion of both the wheels can be derived from the geometry and condition of nonslip. The relationship is as follows.

The following are the parameters of the vehicle.

rho	=	radius of curvature of the path of the center of gravity of the vehicle.
L	=	length of the vehicle.
2b	=	width of the vehicle.
v	=	the longitudinal speed of the vehicle.
omega	=	angular velocity of the vehicle center, w.r.t. the instantaneous center of rotation.

$$vl = v(1 - b/rho);$$
$$vr = v(1 + b/rho);$$

The values generated by the above equations are the exact decimal values and usually fractional numbers, and many times generate recurring values, however, these values may not be achieved always, due to the following limitation of the hardware. The actuation device used in the WMR i.e., stepper motors can take discrete steps only. Hence, it cannot attain all the discrete values of angles generated by the above equations. In such a case the required value would lie between two discrete values achievable by the motor, separated by the step angle of the motor. So one of the ways to solve this problem could be to choose one of the nearest values (usually the nearer) and use it in the vehicle. Rounding off the actual value to the nearest achievable value with respect to the step size can do this. The round-off algorithm can do the rounding off operation, so that the number of steps to be turned by the motors becomes whole numbers.

Determining the Next Step

The system needs a state feedback response to implement closed loop control strategies. But the physical system does not have any state feedback response to determine its current global or local position. So these values have to be determined from the geometry of motion. Since stepper motors are exact actuation devices, the relative displacements of the wheels can be calculated quite

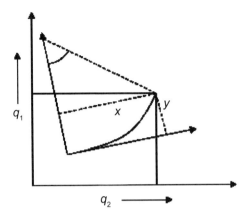

FIGURE 4.36 Determining the next step in a differentially driven WMR.

accurately from the kinematic relationships of the motion. This is of course under the assumption that there is no slipping in the wheels and the stepper motor is strong enough not to miss any step due to insufficient torque. Sufficiently strong stepper motors can be used to ensure that the motor provides enough torque to overcome missing of steps. Thus the stepper motor can also generate very accurate feedback, if modeled correctly. The above WMR is modeled in the following manner to give feedback of its current location.

$$q_1 = q_1 + (x \times \cos q_3 - y \times \sin q_3)$$
$$q_2 = q_2 + (x \times \sin q_3 + y \times \cos q_3)$$
$$q_3 = q_3 + \frac{v_a}{l} \times \tan \delta \times t$$

The above WMR is modeled in the following manner to give feedback. The velocity of the center of the vehicle or the origin of the local coordinate system attached to the vehicle is computed from the corrected velocity, (v_{Rl_{a}},v_{Rr_{a}}) assumed by the wheels. Thus, the values of actual longitudinal velocity v_{a}, and actual steering angle delta_{a} assumed by the vehicle in the previous interval can be computed from the above relations.

Hence, the next position of the vehicle in the global reference frame can be found out as,

$$\omega = \frac{v_a}{p} = \frac{v_a}{l} \tan \delta_a$$

$$x = \frac{a}{\tan \delta} \times \sin\left(\frac{v_a}{l} \times \tan \delta_a \times t\right)$$

$$y = \frac{l}{\tan \delta} \times \left(1 - \cos\left(\frac{v_a}{l} \times \tan \delta_a \times t\right)\right)$$

These values of the position coordinates give the new position and orientation of the vehicle in the global reference frame. Then this point is treated as the instantaneous position of the vehicle, and the entire procedure is repeated until the vehicle reaches the destination point. The vehicle is assumed to follow the trajectory calculated by the geometry of motion flawlessly. The small error occurring due to slip at the wheels can, however, be neglected.

The Algorithms for the Control

Software forms the core of the control system. It comprises the set of algorithms for performing the different functions involved as well as the implementation of these algorithms in the form of a computer program. The entire software system can be considered to consist of three components: stepper motor control software; WMR-specific functions, which include the kinematic model; and graph plotting functions to display the status of motion in real time. The stepper motor control software contains the actual hardware-level functions, which generate the appropriate parallel port signals that interact with the electronic hardware. The language used for programming is C++ compiled under TurboC++ compiler. In the sections that follow, the three software components enumerated above are discussed in detail. The source code for the control software is listed in Appendix I.

Stepper Motor Control Software

As discussed in the previous chapter, driving the stepper motors consist of switching the windings on and off in a particular sequence. The stepper motor control software thus centers on the generation of this sequence of signals. The sequence required at each state of the motor shaft depends on the previous state. The control software thus has a track of the current state of the motor, and it determines the next signal, which is a four-bit sequence, to be generated at the parallel port according to this state by a suitable algorithm. Since the system consists of three stepper motors, the algorithm also includes the selection of the motor to be stepped and the direction in which the step is to be taken.

The stepper motor control software consists essentially of a step function, which takes the motor ID and direction as arguments:

step(motor ID, dir);

This step function is appropriately called by the WMR-specific functions. Figure 4.37 describes the basic stepping algorithm. With four bits for one

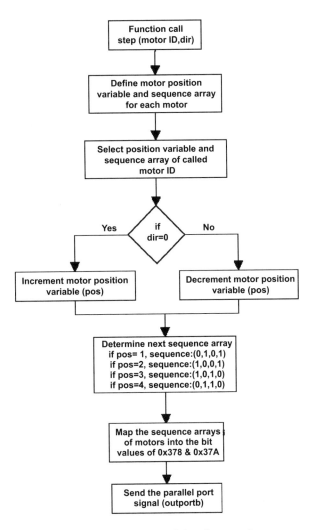

FIGURE 4.37 **Flowchart describing the stepping algorithm.**

motor, the control of three motors requires twelve bits. The parallel port organizes data pins into two sets of 8 and 4 bits each, with port addresses 0x378. Thus port 0x378 handles two motors (for the driving wheels). The step function itself doesn't address the parallel port. After determining the next sequence set for the motor to be stepped, it calls a hardware-level function, which maps the two sequence sets into the bit values of the ports.

outSignal();

The stepper motor control software contains certain additional functions for initializing the motors, for displaying the current position of the motors, and a logging function for logging all the steps and their directions taken by each motor. For initializing the motors, the motors are given the control signals corresponding to the last position in the stepping sequence. This makes sure that stepping takes place as we proceed with the first position onward.

WMR-specific Functions

The kinematic model and the actual functions for controlling the motion of the WMR form the second aspect of the control software. This includes the incorporation of the WMR-specific data such as the geometry, the values of the angle, and the distance traveled in one step of the stepper motor in the form of

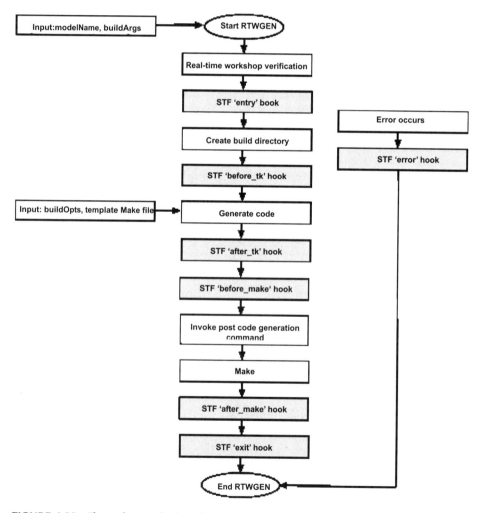

FIGURE 4.38   **Flow of control of motion.**

program variables (l, w, step-distance, step_angle). The kinematic variables—velocity of the center, velocity of the left and the right rear wheels, steering angle, and the coordinates—are also specified here (v, v_Rl, v_Rr, delta, q1, q2, q3). The main functions defined in this part of the control software are the setSteering() and the moveVehicle() functions.

The moveVehicle() function is the function that is actually called by the main program after it calculates the values of v and delta, which are passed as arguments to this function:

moveVehicle (delta, v, t, distance);

The parameter t specifies the time, in seconds, for which the WMR has to travel with the particular values of v and delta. The last argument distance is optional and can be used to specify the distance for which the WMR has to travel instead of specifying the time t. The choice between distance and t is based on the strategy used for the WMR. Figure 4.38 describes the algorithm for this function.

A peculiar problem encountered in the WMR motion function is that the only control over time is by means of the C++ delay() function. The only thing this function is capable of is to suspend the execution of the program for the specified duration. In order to step the rear motors independently, we need to step the motors at appropriate timings in the step-timing array for the individual motors. Since there is no way to execute the stepping sequence simultaneously for the two motors, the problem is overcome by combining the step-timing arrays for the individual motors into a single step-timing array. The stepping sequence is finally executed by calling the step() function for the particular motor at each instant defined in the combined step-timing array.

The second function, setSteering() is called by the moveVehicle() function itself. The moveVehicle() function passes the value of delta as argument to this function:

setSteering (delta);

This function first determines the increment in the value of the steering angle delta with respect to the current value. It then makes the steering motor take the desired number of steps to reach the new value of delta.

Plotting Functions

The plotting portion of the software produces a graphical display of the instantaneous positions of the WMR in a two-dimensional coordinate system. It contains relevant functions for drawing the coordinate system, displaying the position of the WMR at any instant, displaying auxiliary information on the screen such as the instantaneous coordinates (q1, q2, q3), and status of the WMR motion (i.e., steering or moving, etc.). The most important part of the plotting functions is the algorithm for mapping the WMR coordinate system into the screen coordinate system. As opposed to the WMR coordinate system, the screen coordinate system, i.e., the pixel positions, start at the top-left corner of the screen and increase from left to right along the width and from top to bottom along the height. The basic transformations for mapping the WMR coordinate system into the screen coordinate system are enumerated below.

$$q1\_p=q1*q1\_SF+q1\_offset;$$

$$q2\_p=screen\_h-(q2*q2\_SF+q2\_offset);$$

q1_p is the horizontal pixel coordinate corresponding to the q1 axis and q2_p is the vertical pixel coordinate corresponding to the q2 axis. Thus, the WMR coordinates (q1,q2) are mapped into the screen coordinates (q1_p,q2_p). q1_SF and q2_SF are the scaling factors along the q1 and q2 axis respectively. q1_ offset and q2_offset are the horizontal and vertical offsets of the origin of the WMR coordinates from the top-left corner of the screen. The instantaneous position of the WMR is displayed by drawing a point at the appropriate screen coordinates. This is done by means of a plot() function, which carries the transformation and draws a pixel on the screen by using low-level TurboC++ graphics functions:

$$plot(q1,q2,special\_flag).$$

The special_flag argument can be used to specify the color or style of the plotted point. The plot () function is called at appropriate situations by the WMR-specific functions when the stepping sequence of the driving rear motors is executed.

Program Organization

The control software is organized into a set of header files, which correspond to the three components as discussed in the beginning of this chapter. In addition to these three header files, we have the main program, which includes these header files and which contains the strategy for the determination of the values of v and delta. For a simple, predetermined path-following program, the main program serves to simply input the values of v, delta, and t. The control is then transferred to the header files. For automatic tracking, the main program contains the algorithm for calculating the values of v, delta, and t in a loop, which terminates when the desired position has been reached. The general structure of the main program is shown in Figure 4.39.

The header files are included in the beginning of the program so that the main program can access the functions defined in these header files. The functions for initializing the motors and the graphics display are called before the other routines. Then we have the main motion loop of the WMR in which the v and delta are calculated according to the mode of motion i.e., trajectory tracking

```
Include<stepper.h>
Include<wmr.h>
Include<plot.h>
main()
{
initializeMotors();
initializeGFX();
----main motion loop----
calculating u, delta;
moveVehivle(delta,v,t);
---------------------------------
closeGFX();
}
```

FIGURE 4.39

or automatic control. Finally, after the motion loop is executed, the graphics are closed by calling the closeGFX () function.

Running Turtle

The C++ program for the above project is in Appendix II. The program can be run through Turbo C in DOS. Turtle can be run with the parallel port of the computer using the following instructions.

Step 1: Provide turtle with a 12 V regulated power supply. The red and black wires are the positive and negative terminals of the power supply to be fed to Turtle.

Step 2: Connect the parallel port male connector with the female connector in the CPU.

Step 3: Run the executable files in the computer to generate the signals. Working with executable files is described in detail in the following section.

Working with Turtle through an executable file:

Turtle can be run through the executable file, turtle2.exe. The file is located on the CD-ROM in the folder.exe files.

Caution: The files **egavga.bgi** and **egavga.obj** should also be transferred along with the .exe files, in case the files are required to be transferred to another location.

Double-clicking on Turtle2.exe displays the command prompt window shown in Figure 4.41.

The four choices are described here.

L – Choosing this option generates the **log.txt** file in the source folder. This file contains the information about the motion for a detailed analysis later. On each execution a new one displaces the old log file.

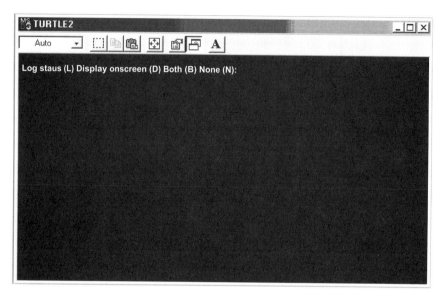

FIGURE 4.40   **Status window for a type of log file generation.**

D – Choosing this option shows the information of the log file on the screen, while the execution takes place. Log file is not generated.

B – Choosing this option shows the information of the log file on the screen. A log file is also generated.

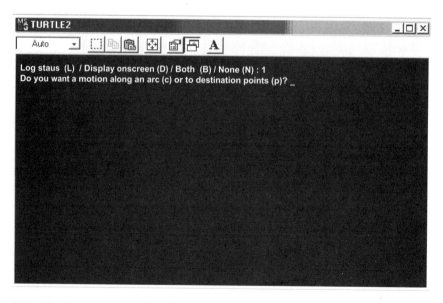

FIGURE 4.41   **Window asking for motion type.**

N – Choosing this option neither shows the information on screen, nor is the log file generated.

Once one of these four choices is selected, the screen in Figure 4.41 appears. It asks the user for the type of motion that is desired, e.g., motion along a circular arc or motion to destination points.

Motion along a circular arc:

If the choice 'c' is entered, the motion takes place along a circular arc. It asks for the following information from the user about the parameters of the path.

Radius – It is the required radius of the curvature of the path that Turtle is required to travel. The value is to be entered in '*millimeters.*'
Velocity – It is the required velocity of travel of the center of location of Turtle. The value is to be entered in '*millimeters/second.*'
Angle – It is the total angle that Turtle has to turn through. The value is to be entered in '*degrees.*'
The screen Figure 4.42 appears when these values are entered.
Once these values are entered, the execution begins and the real-time graphical simulation of the motion in Cartesian coordinates appears on the screen. The screen looks like the Figure 4.43 while the execution takes place.

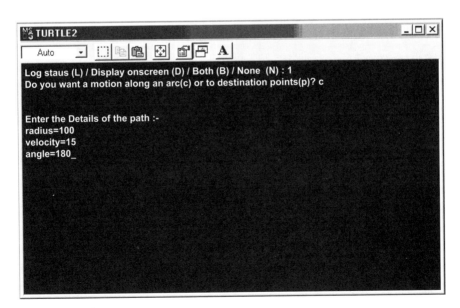

FIGURE 4.42 **Motion along an arc (specifications).**

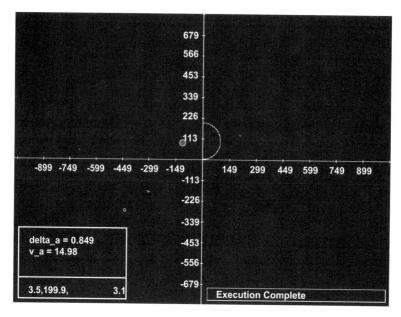

FIGURE 4.43   **Graphical simulation of arc-type motion.**

Motion to destination points:

If the choice 'P' is entered, the motion takes place to the destination points specified by the user. It asks for the following information from the user about the parameters of the path.

x – The desired longitudinal Cartesian coordinate of the destination point.

y – The desired lateral Cartesian coordinate of the destination point.

The screen in Figure 4.44 appears when these values are entered.

The log file generated:

—Interval 1 Step Log—

delta_a=1.107149,   delta_a_increment=3.23646,   motor   F-steps=204, with delay=10ms, in   dir=0

v_a=19.98, v_Rl_a=19.98, v_Rr_a=19.98

Global coordinates of vehicle CoG: [17.74898335.801888,1.11055]

motor Rl - 55 steps

motor Rr - 55 steps

—Interval 2 Step Log—

delta_a=-0.004141,   delta_a_increment=-0.015865,   motor   F-steps=1, with delay=10ms, in   dir=1

v_a=19.98, v_Rl_a=19.98, v_Rr_a=19.98

Global coordinates of vehicle CoG: [35.49796771.603775,1.11055]

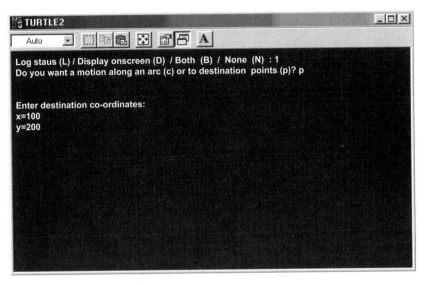

FIGURE 4.44   The screen having the coordinate values entered.

motor Rl - 55 steps
motor Rr - 55 steps

———————————

—Interval 3 Step Log—
delta_a=-0.005293, delta_a_increment=-0.015865, motor F-steps=1,
with delay=10ms, in   dir=1
v_a=19.98, v_Rl_a=19.98, v_Rr_a=19.98
Global coordinates of vehicle CoG: [53.246952107.405663,1.11055]
motor Rl - 55 steps
motor Rr - 55 steps

———————————

—Interval 4 Step Log—
delta_a=-0.007332, delta_a_increment=-0.015865, motor F-steps=1,
with delay=10ms, in   dir=1
v_a=19.98, v_Rl_a=19.98, v_Rr_a=19.98
Global coordinates of vehicle CoG: [70.995934143.20755,1.11055]
motor Rl - 55 steps
motor Rr - 55 steps

———————————

—Interval 5 Step Log—
delta_a=-0.011927, delta_a_increment=-0.03173, motor F-steps=2, with
delay=10ms, in   dir=1
v_a=20.35, v_Rl_a=20.35, v_Rr_a=20.35

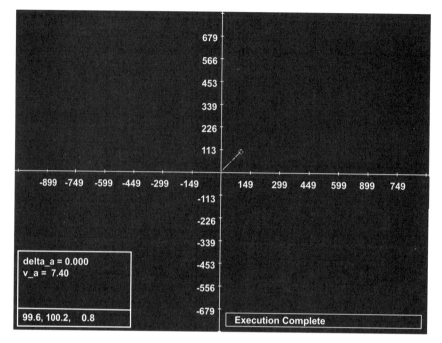

FIGURE 4.45   **Graphical simulation of point-to-point-type motion.**

Global coordinates of vehicle CoG: [89.649818179.381058,1.094685]
    motor Rl - 55 steps
    motor Rr - 55 steps

—Interval 6 Step Log—
delta_a=0.010885,   delta_a_increment=0,   motor   F-steps=0,   with
delay=10ms
    v_a=14.8, v_Rl_a=14.8, v_Rr_a=14.8
Global coordinates of vehicle CoG: [96.433052192.535065,1.094685]
    motor Rl - 20 steps
    motor Rr - 20 steps

—Interval 7 Step Log—
delta_a=0.030359,   delta_a_increment=0,   motor   F-steps=0,   with
delay=10ms
    v_a=7.4, v_Rl_a=7.4, v_Rr_a=7.4
Global coordinates of vehicle CoG: [99.824669199.112061,1.094685]
    motor Rl - 10 steps
    motor Rr - 10 steps

Chapter 5 | KINEMATICS OF ROBOTIC MANIPULATORS

In This Chapter

- Introduction to Robotic Manipulators
- Position and Orientation of Objects in Space
- Forward Kinematics
- Inverse Kinematics

5.1 INTRODUCTION TO ROBOTIC MANIPULATORS

 Most robotic manipulators are strong rigid devices with powerful motors, strong gearing systems, and very accurate models of the dynamic response. For undemanding tasks it is possible to precompute and apply the forces needed to obtain a given velocity. This control is called computed torque control. Alternatively, a high-gain feedback on joint angle control leads to an adequate tracking performance. The important control problem is one of understanding and controlling the manipulator kinematics. Very few robots are regularly pushed to the limit where the dynamic model becomes important since this will lead to greatly reduce operational life and high maintenance costs. In this chapter we consider that part of the manipulator kinematics known as forward kinematics.

213

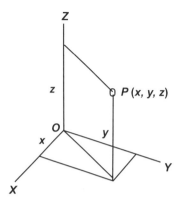

FIGURE 5.1   Position of a point P in
a Cartesian coordinate frame.

5.2   POSITION AND ORIENTATION OF OBJECTS IN SPACE

5.2.1   Object Coordinate Frame: Position, Orientation, and Frames

The manipulator hand's complete information can be specified by position and orientation. The position of a point can be represented in Cartesian space by a set of three orthogonal right-handed axes X, Y, Z, called principal axes, as shown in Figure 5.1. The origin of the principal axes is at O along with three unit vectors along these axes.

The position and orientation pair can be combined together and defined as an entity called frame, which is a set of four vectors, giving position and orientation information.

$$F = \begin{bmatrix} n & s & a & P \\ 0 & 0 & 0 & 1 \end{bmatrix} = \begin{bmatrix} n_x & s_x & a_x & p_x \\ n_y & s_y & a_y & p_y \\ n_z & s_z & a_z & p_z \\ 0 & 0 & 0 & 1 \end{bmatrix} \quad \text{where,} \quad P = \begin{bmatrix} p_x \\ p_y \\ p_z \\ 1 \end{bmatrix}$$

The above equation represents the general representation of a frame. In the above frame, n, s, and a are the unit vectors in the three mutually perpendicular directions, which represent the orientation, and P represents the position vector.

5.2.2   Mapping between Translated Frames

Translation along the z-x axis

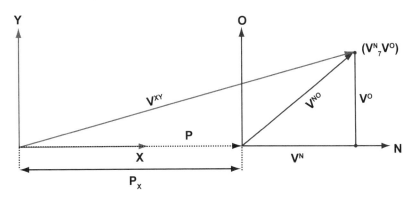

P_x = Distance between XY and NO coordinate frames.

$$\overline{V}^{XY} = \begin{bmatrix} V^X \\ V^Y \end{bmatrix} \quad \overline{V}^{NO} = \begin{bmatrix} V^N \\ V^O \end{bmatrix} \quad \overline{P} = \begin{bmatrix} P_x \\ 0 \end{bmatrix}$$

Translation along the x-axis and y-axis.

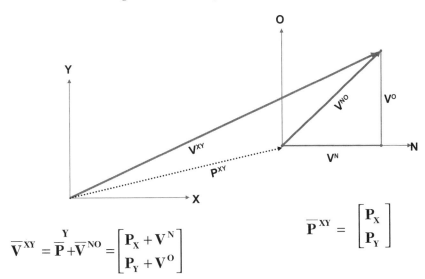

$$\overline{V}^{XY} = \overset{Y}{\overline{P}} + \overline{V}^{NO} = \begin{bmatrix} P_X + V^N \\ P_Y + V^O \end{bmatrix} \qquad \overline{P}^{XY} = \begin{bmatrix} P_X \\ P_Y \end{bmatrix}$$

5.2.3   Mapping between Rotated Frames

Rotation (around the z-axis).

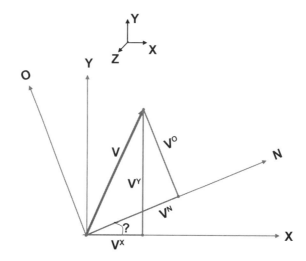

Y = Angle of rotation between the XY and NO coordinate axis.

$$\overline{V}^{XY} = \begin{bmatrix} V^X \\ V^Y \end{bmatrix} \quad \overline{V}^{NO} = \begin{bmatrix} V N \\ V^O \end{bmatrix}$$

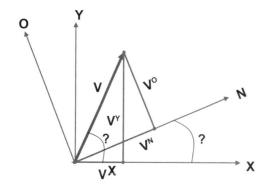

X-Unit vector along the x-axis.
Δ can be considered with respect to the XY coordinates or NO coordinates.

$$\left\| \overline{V}^{XY} \right\| = \left\| \overline{V}^{NO} \right\|$$

$V^X = \left\| \overline{V}^{XY} \right\| \cos \alpha = \left\| \overline{V}^{NO} \right\| \cos \alpha = \overline{V}^{NO} \bullet \overline{x}$

$V^X = (V^N * n + V^O * \overline{o}) \bullet \overline{x}$

$V^X = V^N (\overline{x} \bullet \overline{n}) + V^O (\overline{x} \bullet \overline{o})$

$\qquad = V^N (\cos \theta) + V^O (\cos(\theta + 90))$

(Substituting for VNO using the N and O components of the vector.)

$\qquad = V^N (\cos \theta) - V^O (\sin \theta)$

Similarly...

$$V^Y = \left\|\overline{V}^{NO}\right\| \sin \alpha = \left\|\overline{V}^{NO}\right\| \cos(90 - \alpha) = \overline{V}^{NO} \bullet \overline{y}$$

$$V^Y = (V^N * \overline{n} + V^O * \overline{o}) \bullet \overline{y}$$

$$V^Y = V^N (\overline{y} \bullet \overline{n}) + V^O (\overline{y} \bullet \overline{o})$$

$$= V^N (\cos(90 - \theta)) + V^O (\cos \theta)$$

$$= V^N (\sin \theta) - V^O (\cos \theta).$$

So...

$$V^X = V^N (\cos \theta) - V^O (\sin \theta) \qquad \overline{V}^{XY} = \begin{bmatrix} V^X \\ V^Y \end{bmatrix}$$

$$V^Y = V^N (\sin \theta) + V^O (\cos \theta)$$

Written in matrix form

$$V^{XY} = \begin{bmatrix} V^X \\ V^Y \end{bmatrix} = \begin{bmatrix} \cos \theta & -\sin \theta \\ \sin \theta & \cos \theta \end{bmatrix} \begin{bmatrix} V^N \\ V^O \end{bmatrix} \text{Rotation matrix about the z-axis.}$$

Now generalizing the results for three dimensions, the rotation between two frames about different axes are:

–   Rotation about the x-axis with

$$Rot(x, \theta) = \begin{bmatrix} 1 & 0 & 0 \\ 0 & C\theta & -S\theta \\ 0 & S\theta & C\theta \end{bmatrix}.$$

–   Rotation about the y-axis with

$$Rot(y, \theta) = \begin{bmatrix} C\theta & 0 & S\theta \\ 0 & 1 & 0 \\ -S\theta & 0 & C\theta \end{bmatrix}.$$

–   Rotation about the z-axis with θ

$$Rot(z, \theta) = \begin{bmatrix} C\theta & -S\theta & 0 \\ S\theta & C\theta & 0 \\ 0 & 0 & 1 \end{bmatrix}.$$

Where,

$C\,\theta$= Cos
S = Sin

a general rotation between any two frames can be represented as:

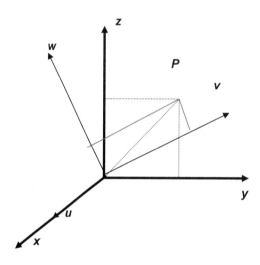

$$P_{xyz=}\begin{bmatrix} p_x \\ p_y \\ p_z \end{bmatrix} = \begin{bmatrix} i_x \cdot i_u & i_x \cdot j_v & i_x \cdot k_w \\ j_y \cdot i_u & j_y \cdot j_v & j_y \cdot k_w \\ k_z \cdot i_u & k_z \cdot j_v & k_z \cdot k_w \end{bmatrix} \begin{bmatrix} p_u \\ p_v \\ p_w \end{bmatrix} = RP_{uvw}$$

5.2.4   Mapping between Rotated and Translated Frames

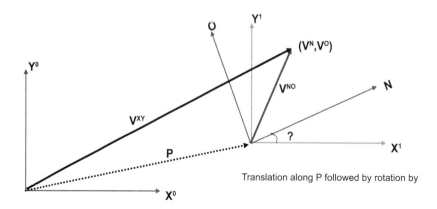

Translation along P followed by rotation by

$$\mathbf{V}^{XY} = \begin{bmatrix} \mathbf{V}^X \\ \mathbf{V}^Y \end{bmatrix} = \begin{bmatrix} \mathbf{P}_x \\ \mathbf{P}_y \end{bmatrix} + \begin{bmatrix} \cos\theta & -\sin\theta \\ \sin\theta & \cos\theta \end{bmatrix} \begin{bmatrix} \mathbf{V}^N \\ \mathbf{V}^O \end{bmatrix}$$

(Note: Px, Py are relative to the original coordinate frame. Translation followed by rotation is different than rotation followed by translation.)

In other words, knowing the coordinates of a point (VN, VO) in some coordinate frame (NO) you can find the position of that point relative to your original coordinate frame (XOYO).

5.2.5 Homogeneous Representation

Putting it all into a matrix.

$$\mathbf{V}^{XY} = \begin{bmatrix} \mathbf{V}^X \\ \mathbf{V}^Y \end{bmatrix} = \begin{bmatrix} \mathbf{P}_x \\ \mathbf{P}_y \end{bmatrix} + \begin{bmatrix} \cos\theta & -\sin\theta \\ \sin\theta & \cos\theta \end{bmatrix} \begin{bmatrix} \mathbf{V}^N \\ \mathbf{V}^O \end{bmatrix}$$

What we found by doing a translation and a rotation.

$$= \begin{bmatrix} \mathbf{V}^X \\ \mathbf{V}^Y \\ 1 \end{bmatrix} = \begin{bmatrix} \mathbf{P}_x \\ \mathbf{P}_y \\ 1 \end{bmatrix} + \begin{bmatrix} \cos\theta & -\sin\theta & 0 \\ \sin\theta & \cos\theta & 0 \\ 0 & 0 & 0 \end{bmatrix} \begin{bmatrix} \mathbf{V}^N \\ \mathbf{V}^O \\ 1 \end{bmatrix}$$

Padding with 0s and 1s.

$$= \begin{bmatrix} \mathbf{V}^X \\ \mathbf{V}^Y \\ 1 \end{bmatrix} = \begin{bmatrix} \cos\theta & -\sin\theta & \mathbf{P}_x \\ \sin\theta & \cos\theta & \mathbf{P}_y \\ 0 & 0 & 1 \end{bmatrix} \begin{bmatrix} \mathbf{V}^N \\ \mathbf{V}^O \\ 1 \end{bmatrix}$$

Simplifying into a matrix form.

$$H = \begin{bmatrix} \mathbf{V}^X \\ \mathbf{V}^Y \\ 1 \end{bmatrix} = \begin{bmatrix} \cos\theta & -\sin\theta & \mathbf{P}_x \\ \sin\theta & \cos\theta & \mathbf{P}_y \\ 0 & 0 & 1 \end{bmatrix} \begin{bmatrix} \mathbf{V}^N \\ \mathbf{V}^O \\ 1 \end{bmatrix}$$

Homogeneous matrix for a translation in the XY plane, followed by a rotation around the z-axis.

The coordinate transformation from frame B to frame A can be represented as:

$$^A r^P = {}^A R_B \, {}^B r^P + {}^A r^{o'}$$

$$\begin{bmatrix} {}^A r^P \\ 1 \end{bmatrix} = \begin{bmatrix} {}^A R_B & {}^A r^{o'} \\ 0_{1\times3} & 1 \end{bmatrix} \begin{bmatrix} {}^B r^P \\ 1 \end{bmatrix}.$$

Homogeneous transformation matrix

$$^A T_B = \begin{bmatrix} {}^A R_B & {}^A r^{o'} \\ 0_{1\times3} & 1 \end{bmatrix} = \begin{bmatrix} R_{3\times3} & P_{3\times1} \\ 0 & 1 \end{bmatrix}.$$

5.3   FORWARD KINEMATICS

In this section, we will develop the forward or configuration kinematic equations for rigid robots. The forward kinematics problem is concerned with the relationship between the individual joints of the robot manipulator and the position and orientation of the tool or end-effector. Stated more formally, the forward kinematics problem is to determine the position and orientation of the end-effector, given the values for the joint variables of the robot. The joint variables are the angles between the links in the case of revolute or rotational joints, and the link extension in the case of prismatic or sliding joints. The forward kinematics problem is to be contrasted with the inverse kinematics problem, which will be studied in the next section, and which is concerned with determining values for the joint variables that achieve a desired position and orientation for the end-effector of the robot.

5.3.1   Notations and Description of Links and Joints

A robot manipulator is composed of a set of links connected together by various joints. The joints can either be very simple, such as a revolute joint or a prismatic joint, or they can be more complex, such as a ball and socket joint. A revolute joint is like a hinge and allows a relative rotation about a single axis, and a prismatic joint permits a linear motion along a single axis, namely an extension or retraction. The difference between the two situations is that, in the first instance, the joint has only a single degree of freedom of motion: the angle of rotation in the case of a revolute joint, and the amount of linear displacement in the case of a prismatic joint. In contrast, a ball and socket joint has two degrees of freedom. With the assumption that each joint has a single degree of freedom, the action of each joint can be described by a single real number: the angle of rotation in the case of a revolute joint or the displacement in the case of a prismatic joint. The objective of forward kinematic analysis is to determine the cumulative effect of the entire set of joint variables. In this section we will develop a set of conventions that provide a systematic procedure for performing this analysis. It is, of course, possible to carry out forward kinematics analysis even without respecting these conventions. However, the kinematic analysis of an n-link manipulator can be extremely complex and the conventions introduced below simplify the analysis considerably. Moreover, they give rise to a universal language with which robot engineers can communicate.

The following steps are followed to determine the forward kinematics of a robotic manipulator.

1. Attach an inertial frame to the robot base.
2. Attach frames to links, including the end-effector.
3. Determine the homogenous transformation between each frame.
4. Apply the set of transforms sequentially to obtain a final overall transform.

However, there is a standard way to carryout these steps for robot manipulators; it was introduced by Denavit and Hartenberg in1955 (J. Denavit and R.S. Hartenberg, "A Kinematic Notation for Lower-Pair Mechanisms Based on Matrices," Journal of Applied Mechanics, pp. 215–221, June 1955.) The key element of their work was providing a standard means of describing the geometry of any manipulator, so that step 2 above becomes obvious. A robotic manipulator is a chain of rigid links attached via a series of joints. Given below is a list of possible joint configurations.

1. Revolute joints: Are comprised of a single fixed axis of rotation.
2. Prismatic joints: Are comprised of a single linear axis of movement.
3. Cylindrical joints: Comprise two degrees of movement, revolute around an axis and linear along the same axis.
4. Planar joints: Comprise two degrees of movement, both linear, lying in a fixed plane (A gantry-type configuration).
5. Spherical joints: Comprise two degrees of movement, both revolute, around a fixed point (A ball joint conûguration).
6. Screw joints: Comprised of a single degree of movement combining rotation and linear displacement in a fixed ratio.

However, the last 4 joint configurations can be modeled as a degenerate concatenation of the first two basic joint types.

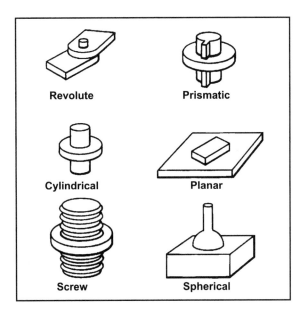

FIGURE 5.2 **Some possible joint configurations.**

5.3.2   Denavit-Hartenberg Notation

Denavit-Hartenberg notation looks at a robot manipulator as a set of serially attached links connected by joints. Only joints with a single degree of freedom are
considered. Joints of a higher order can be modeled as a combination of single
dof joints. Only prismatic and revolute joints are considered. All other joints are
modeled as combinations of these fundamental two joints. The links and joints
are numbered starting from the immobile base of the robot, referred to as link
0, continuing along the serial chain in a logical fashion. The first joint, connecting the immobile base to the first moving link is labeled joint1, while the first
movable link is link1. Numbering continues in a logical fashion. The geometrical
configuration of the manipulator can be described as a 4-tuple, with 2 elements
of the tuple describing the geometry of a link relative to the previous link.

> A: Link length
> α : Link twist

And the other 2 elements describing the linear and revolute offset of the link:

> d : Link offset
> θ : Joint angle

Let's look at the details of these parameters for link i-1 and joint i of the chain.

Link length, a_{i-1}: Consider the shortest distance between the axis of link i-1
and link i in R3. This distance is realized along the vector mutually perpendicular
to each axis and connecting the two axes. The length of this vector is the link

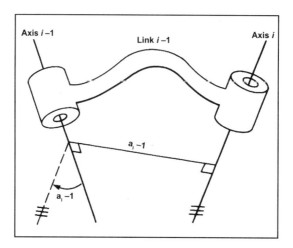

FIGURE 5.3   Representation of link length.

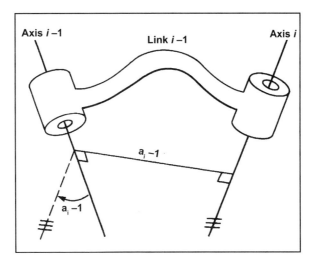

FIGURE 5.4   Representation of link twist.

length $a_{i\text{-}1}$. Note that link length need not be measured along a line contained in the physical structure of the link. Although only the scalar link length is needed in the mathematical formulation of joint transformations, the vector direction between joint axes is also important in understanding the geometry of a robotic manipulator. Thus, we use the terminology of link length both as a scalar denoting the distance between links and as a vector $a_{i\text{-}1}$ direction that points from the axis of joint (i-1) to joint i and such that $a_{i\text{-}1} = |a_{i\text{-}1}|$.

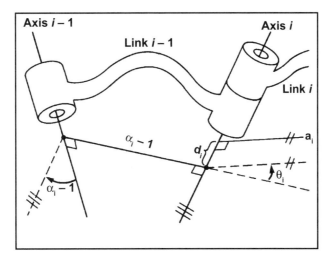

FIGURE 5.5   Representation of link offset.

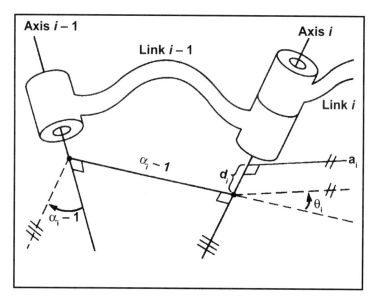

FIGURE 5.6   **Representation of joint angle.**

Link Twist, α_{i-1}: Consider the plane orthogonal to the link length a_{i-1}. Both axis vectors of joint i-1 and I lie in this plane. Project the axes vectors of joints I and i-1 onto this plane. The link twist is the angle measured from joint axis i-1 to joint axis i in the right-hand sense around the link length a_{i-1}. Direction of a_{i-1} taken as from axis i-1 to i. This is to say that α_{i-1} will be positive when the link twist (by the right-hand rule) is in the positive direction of a_{i-1}.

Link Offset, d_i: On the joint axis of joint I consider the two points at which the link lengths a_{i-1} and ai are attached. The distance between these points is the link offset, measured positive from the a_{i-1} to ai connection points.

Joint Angle, θ_i: Consider a plane orthogonal to the joint axis i. By construction, both link length vectors a_{i-1} and ai lie in this plane. The joint angle is calculated as the clockwise angle that the link length a_{i-1} must be rotated to be colinear with link length ai. This corresponds to the right-hand rule of a rotation of link length a_{i-1} about the directed joint axis.

5.3.3   First and Last Links in the Chain

Certain parameters in the first and last links in a chain are automatically specified, or can be arbitrarily specified by convention:

$$a_0 = 0 = a_n$$
$$\alpha_0 = 0 = \theta_n$$

If joint 1 (resp. joint n) is revolute:

1. The zero position for d1 (resp. d_n) can be chosen arbitrarily.
2. The link offset is set to zero $d_1 = 0$ (resp. $d_n = 0$).

If joint 1 (resp. joint n) is prismatic:

1. The zero position for d_1 (resp. d_n) can be chosen arbitrarily.
2. The joint angle is set to zero $\theta_1 = 0$ (resp. $\theta_n = 0$).

5.3.4 Summary: D.H. Parameters

The four parameters are:

a_i Link length: Displacement of joint axis I from joint axis i-1.
α_i Link twist: Twist of axis i with respect to axis i-1.
d_i Link offset: Linear displacement of the joint i along the axis of joint i.
θ_i Joint angle: Rotational displacement of the joint i around the axis of joint i.

NOTE

- For a revolute joint, link offset is fixed and joint angle is a controlled variable.
- For a prismatic joint, joint angle is fixed and link offset is a controlled variable.

The first two parameters, link length and link twist, are always fixed parameters. So, for any robot with n single-dof revolute or prismatic joints, there will be:

- Three n fixed parameters, termed the link parameters. The link parameters describe the fixed kinematics of the mechanism.
- N controlled parameters (one for each joint), termed the joint variables.

For example, for a six-jointed robot with all revolute joints (anthropomorphic arm) the link parameters are (a_i, α_i, d_i) for i=1,..., 6, a set of 18 numbers. Applying the conventions for the zero link and last link of a robotic manipulator

$$(a_0, \alpha_0) = (0, 0) = (a_n, \alpha_n).$$

(The world frame is taken as fixed at the center of the first joint.)

Since the first link is revolute $d_1 = 0$.
Since the last link is revolute $d_n = 0$.

Thus, the geometry of an anthropomorphic robotic manipulator is speciûed by 14 numbers a_1 and $\bullet_1$ along with (a_i, α_i, d_i) for I = 2,..., 5. The joint variables are $(\theta_1,..., \theta_6)$.

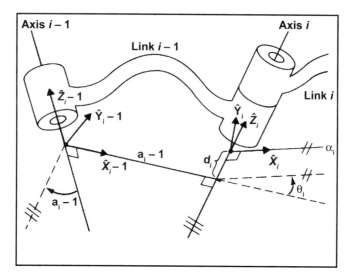

FIGURE 5.7   Schematic representation of a link.

5.3.5   Kinematic Modeling Using D-H Notations

With the machinery of the Denavit-Hartenberg notation available, the process of attaching frames to links for the purpose of determining the manipulator's forward kinematics is relatively straight forward.

A frame is attached to each link of the robot manipulator. The frame attached to link I is denoted as {i}.

1. The origin of the frame for link I is placed at the intersection of the joint axis of link I with the vector direction ai (connecting link axis i with link axis i + 1).
2. The direction Z_i is chosen in the direction of the link axis.
3. The direction X_i is chosen to lie along the vector ai connecting link axis i to axis i + 1. (Note that choosing the direction of X_i is equivalent to choosing the direction in which the twist ϵ_i is measured.)
4. The direction Y_i is fixed by the choice of X_i and Z_i and the right-hand rule, $Y_i = Z_i \times X_i$.

5.3.6   Special Cases

Base Link: The base frame (orlink0) is the effective inertial frame for the manipulator kinematics. Choose this inertial frame such that it is coincident with link frame 1 when the robot is its zeroed position.

Thus,

$$a_0 = 0, \ \alpha_0 = 0$$

and

$d_1 = 0$ if joint 1 is revolute or $\theta_1 = 0$ if joint 1 is prismatic.

Final Link: Again, choose the frame for link n coincident with the frame for link n-1 in the robot zeroed position. Thus, again

$a_n = 0$, $\alpha_n = 0$

and

$d_n = 0$ if joint 1 is revolute or $\alpha_n = 0$ if joint 1 is prismatic.

Link i:

1. If the joint length $a_i = 0$ is zero (i.e., intersecting joint axes), choose X_i to be orthogonal to the plane spanned by $\{Z_i, Z_{i+1}\}$.
2. If $\{Z_i, Z_{i+1}\}$ are collinear, then the only non trivial arrangements of joints is either prismatic/revolute or revolute/prismatic, i.e., a cylindrical joint.
3. The joint angle $\theta_i = 0$ in the zeroed position of the robot.

Link Parameters in Terms of Attached Frames

a_i : The distance from Z_i to Z_{i+1} measured along X_i.
a_i : The vector distance $a_i X_i$.
α_i : The angle between Z_i and Z_{i+1} measured about the axis X_i.
d_i : The distance from X_{i-1} to X_i measured along Z_i.
θ_i : The angle between X_{i-1} and X_i measured about the axis Z_i.

Note that the conventions are usually chosen such that ai > 0 is a consequence of the choices made. Since ai is a distance, it is generally written as a positive number even if X_i is chosen in the negative direction.

Summary of Link Frame Attachment Procedure

1. Identify the joint axes and imagine (or draw) infinite lines along them. For steps 2 through 5 below, consider two neighboring axes (i and i+1).
2. Identify the common perpendicular, or point of intersection, between the neighboring axes. At the point of intersection, or at the point where the common perpendicular meets the *i*th axis, assign the link frame origin.
3. Assign the Z_i axis pointing along the *i*th joint axis.
4. Assign the X_i axis pointing along the common perpendicular, or if the axes intersect, assign X_i to be normal to the plane containing the two axes.
5. Assign the Y_i axis to complete a right-hand coordinate system.
6. Assign the {0} frame to match the {1} frame when the first joint variable is zero. For the {N} frame choose an origin location and XN direction freely, but generally, so as to cause as many linkage parameters as possible to become zero.

Note that the frame attachment convention above does not result in a unique attachment of frames. For example, the Z_i axis can be attached in either direction of the frame axis. This is not a problem—we end up with the same answer.

5.3.7   Forward Kinematics of a Manipulator

We now have frames attached to each link of the manipulator, including an inertial frame at the base of the robot.

We know we want to solve for ${}^0_N T$ as per

$$ {}^0_N T = {}^0_1 T \, {}^1_2 T \dots {}^{i-1}_i T \dots {}^{N-1}_N T. $$

However, we need the 4x4 homogeneous transformation matrices corresponding to ${}^{i-1}_i T$, i=1...N, and for a general robotic mechanism. These are difficult to write down from inspection.

Recall that the rigid body transformation between any two links

$$ {}^{i-1}_i T : \{i\} \quad \rightarrow \quad \{i-1\} $$

depends on the three link parameters a_{i-1} and z_{i-1} (and either θ_i or d_i depending on whether the joint is prismatic or revolute). The joint variable (either d_i or θ_i) is actuated so that the transformation from frame {i} to frame {i-1} can be written down as

$$ {}^{i-1}_i T := {}^{i-1}_i T_{(a_{i-1}, \alpha_{i-1}, d_i)} \, (\theta_i) , \qquad {}^{i-1}_i T := {}^{i-1}_i T_{(a_{i-1}, \alpha_{i-1}, \theta_i)} (d_i) $$

for a revolute (resp. prismatic) *i*th joint.

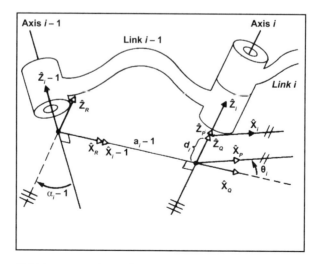

FIGURE 5.8   Schematic representation of a link.

We can write down the 4x4 homogeneous transformation matrix representing $_i^{i-1}T$ by inspection by introducing three other frames to each link. Denote these frames as {P},{R}, and {Q}.

Computing the Transformation from Link i to Link i-1

For each link i-1 assign the three intermediate frames of reference {P}, {R}, and {Q} by:

1. Frame {R} is made coincident with frame {i-1} except for a rotation about the joint i-1 axis by α_{i-1}.
2. Frame {Q} is given the same orientation as {R}, but is translated along X_{i-1} by a_{i-1} so that its origin lies on the axis of joint i.
3. Frame {P} is made coincident with frame {Q} except for a rotation about the joint I axis by θ_i. It then goes without saying that frame {P} and frame {i} differ only by a translation di.

We then may write $_i^{i-1}T = _R^{i-1}T(\alpha_{i-1})_Q^R T(a_{i-1})_P^Q T(\theta_i)_i^P T(d_i)$.

Note that each transformation depends on a single parameter, so we can easily write down each element on the RHS.

Expanding out gives the full expression for the link transformation:

$$_i^{i-1}T = \begin{pmatrix} C\theta_i & -S\theta_i & 0 & a_{i-1} \\ S\theta_i C\alpha_{i-1} & C\theta_i C\alpha_{i-1} & -S\alpha_{i-1} & -S\alpha_{i-1}d_i \\ S\theta_i S\alpha_{i-1} & C\theta_i S\alpha_{i-1} & C\alpha_{i-1} & C\alpha_{i-1}d_i \\ 0 & 0 & 0 & 1 \end{pmatrix}.$$

Actuator Space, Joint Space, and Cartesian Space

The position and orientation of the end-effector and its pose can be parameterized in a number of different coordinates paces.

For an n-degree-of-freedom (dof) robot, we need generally require n parameters to describe the end-effectors pose.

1. Cartesian space is standard Euclidean position along with orientation information. The pose of the end-effector in Cartesian space is given by the homogeneous transformation $_N^0 T$.
2. Joint space is the parameterization given by the set of joint variables. For example, for a SCAR A robot with a single degree of freedom in the wrist $(\theta_1, \theta_2, d_3, \theta_4)$.
3. Actuator space is associated with the mechanism used to actuate a joint. Thus, in certain situations, a linear actuator (say, a hydraulic cylinder) is used to actuate a revolute joint. The actuator space has each of its coordinate axes defined by one of the actuator variables.

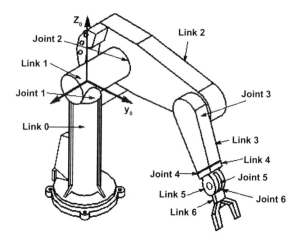

FIGURE 5.9　**PUMA industrial robot with 6 revolute joints.**

We have seen that the mapping from joint space into Cartesian space is accomplished via the homogeneous transformation ${}^0_N T = {}^0_1 T(\theta_1) {}^1_2 T(\theta_2)....{}^{N-1}_N T(\theta_N)$ and is known as the forward kinematics of the manipulator.

The reverse mapping ${}^0_N T \rightarrow (\theta_1........\theta_N)$ is known as the inverse kinematics. We will cover this in the next section.

5.3.8　Examples of Forward Kinematics

The solution of the direct kinematics problem is now explained for an industrial robot with 6 joints, which is the PUMA 562 robot. Figure 5.9 shows the robot with its joints and links. Figure 5.10 presents the robot with its different frame locations.

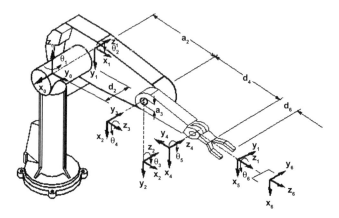

FIGURE 5.10　**DH-frame assignment for the PUMA robot.**

The link frame assignment for the PUMA 562 is presented according to DH-notation in Figure 5.10.

The location of DH-frames and DH-parameters is shown in Figure 5.11, where the 'zero' position for the PUMA joints is assumed.

Transformations between all neighboring frames are given by the following DH-transformations:

$$
{}^{0}T_{1} = \begin{pmatrix} \cos\theta_1 & 0 & -\sin\theta_1 & 0 \\ \sin\theta_1 & 0 & \cos\theta_1 & 0 \\ 0 & -1 & 0 & 0 \\ 0 & 0 & 0 & 0 \end{pmatrix}
$$

$$
{}^{1}T_{2} = \begin{pmatrix} \cos\theta_2 & -\sin\theta_2 & 0 & a_2\cos\theta_2 \\ \sin\theta_2 & \cos\theta_2 & 0 & a_2\sin\theta_2 \\ 0 & 0 & 0 & d_2 \\ 0 & 0 & 0 & 0 \end{pmatrix}
$$

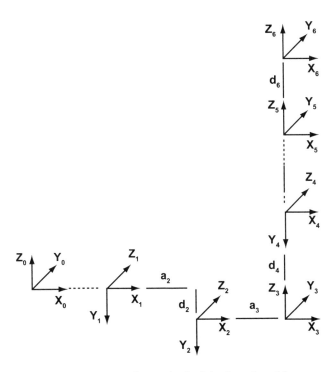

FIGURE 5.11   **PUMA frames in the joint 'zero' position.**

$$
{}^2T_3 = \begin{pmatrix} \cos\theta_3 & 0 & -\sin\theta_2 & a_3\cos\theta_3 \\ \sin\theta_3 & 0 & \cos\theta_2 & a_3\sin\theta_3 \\ 0 & 1 & 0 & d_2 \\ 0 & 0 & 0 & 1 \end{pmatrix}
$$

$$
{}^3T_4 = \begin{pmatrix} \cos\theta_4 & 0 & -\sin\theta_4 & 0 \\ \sin\theta_4 & 0 & \cos\theta_4 & 0 \\ 0 & -1 & d_4 & 0 \\ 0 & 0 & 0 & 1 \end{pmatrix}
$$

$$
{}^4T_5 = \begin{pmatrix} \cos\theta_5 & 0 & -\sin\theta_5 & 0 \\ \sin\theta_5 & 0 & -\cos\theta_2 & 0 \\ 0 & 1 & 0 & 0 \\ 0 & 0 & 0 & 1 \end{pmatrix}
$$

$$
{}^5T_6 = \begin{pmatrix} \cos\theta_6 & -\sin\theta_6 & 0 & 0 \\ \sin\theta_6 & \cos\theta_6 & 0 & 0 \\ 0 & 0 & 0 & d_6 \\ 0 & 0 & 0 & 1 \end{pmatrix}
$$

Multiplication of these matrices leads to the complete transformation from tool frame toward base frame, which solves the direct kinematics problems:

$$
{}^0T_6 = {}^0T_1\,{}^1T_2\,{}^2T_3\,{}^3T_4\,{}^4T_5\,{}^5T_6 = \begin{pmatrix} n_x & s_x & a_x & p_x \\ n_y & s_y & a_y & p_y \\ n_z & s_z & a_z & p_z \\ 0 & 0 & 0 & 1 \end{pmatrix}
$$

$$
n_x = c_1(c_{23}(c_4c_5c_6 - s_4s_6) - s_{23}s_5c_6) - s_1(s_4c_5c_6 + c_4s_6)
$$

$$
n_y = s_1(c_{23}(c_4c_5c_6 - s_4s_6) - s_{23}s_5c_6) - c_1(s_4c_5c_6 + c_4s_6)
$$

$$
n_z = -s_{23}(c_4c_5c_6 - s_4s_5) - c_{23}s_5c_6
$$

$$
s_x = c_1(c_{23}(c_4c_5s_6 - s_4c_6) - s_{23}s_5s_6) - s_1(-s_4c_5s_6 + c_4s_6)
$$

$$
s_y = s_1(-c_{23}(c_4c_5s_6 - s_4c_6) + s_{23}s_5s_6) + c_1(-s_4c_5s_6 + c_4c_6)
$$

$$
s_z = s_{23}(c_4c_5s_6 + s_4s_6) - c_{23}s_5s_6
$$

$$a_x = c_1(c_{23}c_4s_5 + s_{23}c_5) - s_2s_4s_5$$
$$a_y = S_1(c_{23}c_4s_5 + s_{23}c_5) + c_1s_4s_5$$
$$a_z = -s_{23}c_4s_5 + c_{23}c_5$$

The following abbreviations have been used here:

$$s_i = \sin\theta_i$$
$$c_i = \cos\theta_i$$
$$s_{ij} = \sin(\theta_i + \theta_j)$$
$$c_{ij} = \cos(\theta_i + \theta_j)$$

5.4   INVERSE KINEMATICS

Inverse kinematics is the determination of all possible and feasible sets of joint variables, which would achieve the specified positions and orientations of the manipulator's end-effector with respect to the base frame. In practice, a robot manipulator control requires knowledge of the end-effector position and orientation for the instantaneous location of each joint as well as knowledge of the joint displacements required to place the end-effector in a new location. Many industrial applications such as welding and certain types of assembly operations require that a specific path should be negotiated by the end-effector. To achieve this, it is necessary to find the corresponding motion of each joint, which will produce the desired tip motion. This is a typical case of inverse kinematic application.

5.4.1   Workspace

The workspace of a manipulator is defined as the volume of space in which the manipulator is able to locate its end-effector. The work space gets specified by the existence or nonexistence of solutions to the inverse problem. The region that can be reached by the origin of the end-effector frame with at least one orientation is called the reachable workspace (RWS). If a point in workspace can be reached only in one orientation, the manipulatoribility of the end-effector is very poor and it is not possible to do any practical work satisfactorily with just one fixed orientation. It is, therefore, necessary to look for the points in workspace that can be reached in more than one orientation. The space where the end-effector can reach every point from all orientations is called dexterous workspace (DWS). It is obvious that the dexterous workspace is either smaller (subset) or the same as the reachable workspace.

As an example, consider a two-link, nontrivial (2-DOF)-planar manipulator having link lengths L_1 and L_2. The RWS for this manipulator is plane annular space with radii $r_1 = L_1 + L_2$ and $r_2 = |L_1 - L_2|$. The DWS for this case is null. Inside the RSW there are two possible orientations of the end-effector for a given position, while on the boundaries of the RWS the end-effector has only one orientation. For the special case of $L_1 = L_2$, the RWS is a circular area and DWS is a point at the center. It can be shown that for a 3-DOF redundant planar manipulator having link-lengths L_1, L_2, and L_3 with $(L_1 + L_2) > L_3$, the RWS is a circle of radius $(L_1 + L_2 + L_3)$, while the DWS is a circle of radius $(L_1 + L_2 - L_3)$.

The reachable workspace of an n-DOF manipulator is the geometric locus of the points that can be achieved by the origin of the end-effector frame as determined by the position vector of the direct kinematic model. To locate the tool point or end-effector at n arbitrary position with an arbitrary orientation in 3D space, a minimum of 6-DOF are required.

The dexterous workspace may almost approach the reachable workspace. The manipulator work space is characterized by the mechanical joint limits in addition to the configuration and the number of degrees of freedom of the manipulator. In practice, the joint range of revolute motion is much less than 360 degrees for the revolute joints and is severely limited for prismatic joints, due to mechanical constraints. This limitation greatly reduces the workspace of the manipulator and the shape of workspace may not be similar to the ideal case.

To understand the effect of mechanical joint limits on the workspace, consider the 2-DOF planar manipulator with $L_1 > L_2$ and joint limits on θ_1 and θ_2

$$-60^0 \leq q_1 \leq 60^0$$
$$-100^0 \leq q_2 \leq 100^0.$$

For these joint limits, considering $\theta_1 = \theta_2 = 0$ as home position, the annular workspace gets severely limited. The work space, obtained geometrically, is not annular any more, rather it has a complex shape.

Thus, the factors that decide the workspace of a manipulator apart from the number of degrees of freedom are the manipulator's configuration, link lengths, and the allowed range of joint motions.

5.4.2 Solvability

Inverse kinematics is complex because solutions are found for nonlinier simultaneous equations, involving transcendental (harmonic sine and cosine) functions. The number of simultaneous equations is also generally more than the number of unknowns, making some of the equations mutually dependent. These conditions lead to the possibility of multiple solutions or nonexistence of any solution for the given end-effector position and orientation. The existence of solutions, multiple solutions, and methods of solutions are discussed in the following sections.

There are two approaches to the solution of the inverse problem, closed-form solutions and numerical solutions. In the closed-form solution, joint displacements are determined as explicit functions of the position and orientation of the end-effector. In numerical methods, iterative algorithms such as the Newton-Raphson method are used. The numerical methods are computationally intensive and by nature slower compared to closed-form methods. Iterative solutions do not guarantee convergence to the correct solution in singular and degenerate cases.

The closed form in the present context means a solution method based on analytical algebraic or kinematic approach, giving expressions for solving unknown joint displacements. The closed-form solutions may not be possible for all kinds of structures. A sufficient but not necessary condition for a 6-DOF manipulator to possess closed-form solutions is that either its three consecutive joint axes intersect or its three consecutive joint axes are parallel. The kinematic equations under either of these conditions can be reduced to algebraic equations of a degree less than or equal to four, for which closed-form solutions exist. Almost every industrial manipulator today satisfies one of these conditions so that closed-form solutions may be obtained. Manipulator arms with other kinematic structures may be solvable by analytical methods.

5.4.3   Closed form Solutions

Twelve equations, out of which only six are independent, are obtained by equating the elements of the manipulator transformation matrix with end-effector configuration matrix T. At the same time, only six of the twelve elements of T

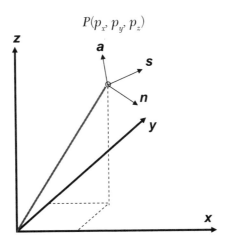

FIGURE 5.12   A frame in Cartesian space.

specified by the end-effector position and orientation are independent. For a manipulator with more than six DOF, the number of independent equations may also be fewer than six. Several approaches such as, inverse transform, screw algebra, and kinematic approach, and so on, can be used for solving these equations but none of them in general so as to solve the equations for every manipulator. A composite approach based on direct inspection, algebra, and inverse transform is presented here, which can be used to solve the inverse equations for a class of simple manipulators.

$$F = \begin{bmatrix} n & s & a & P \\ 0 & 0 & 0 & 1 \end{bmatrix} = \begin{bmatrix} n_x & s_x & a_x & p_x \\ n_y & s_y & a_y & p_y \\ n_z & s_z & a_z & p_z \\ 0 & 0 & 0 & 1 \end{bmatrix} \quad \text{Where} \quad P = \begin{bmatrix} p_x \\ p_y \\ p_z \\ 1 \end{bmatrix}$$

Another useful technique to reduce the complexity is dividing the problem into two smaller parts—the inverse kinematics of the arm and the inverse kinematics of the wrist. The solutions of the arm and the wrist each with, say, 3-DOF are obtained separately. These solutions are combined by coinciding the arm end-point frame with the wrist-based frame to get the total manipulator solution.

5.4.4   Algebraic vs. Geometric Solution

Inverse Kinematics of a Two-link Manipulator

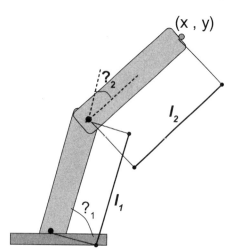

Given: l_1, l_2, x, y

Find: U_1, U_2

Redundancy: A unique solution to this problem does not exist. Notice, that using the "given" two solutions are possible. Sometimes no solution is possible.

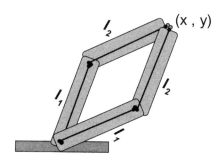

The Geometric Solution

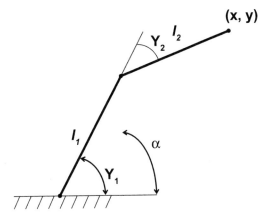

Using the law of cosines:

$$\frac{\sin B}{b} = \frac{\sin C}{c}$$

$$\frac{\sin \overline{\theta}_1}{l_2} = \frac{\sin(180-\theta_2)}{\sqrt{x^2+y^2}} = \frac{\sin(\theta_2)}{\sqrt{x^2+y^2}}$$

$$\theta_1 = \overline{\theta}_1 + a$$

$$\alpha = \arctan 2\left(\frac{y}{x}\right)$$

Using the law of cosines:

$$c^2 = a^2 + b^2 - 2ab\cos C$$

$$(x^2+y^2) = l_1^2 + l_2^2 - 2l_1l_2 \cos(180-\theta_2)$$

$$\cos(180-\theta_2) = -\cos(\theta_2)$$

$$\cos(\theta_2) = \frac{x^2+y^2-l_1^2-l_2^2}{2l_1l_2}$$

$$\theta_2 = \arccos\left(\frac{x^2+y^2-l_1^2-l_2^2}{2l_2l_2}\right)$$

Redundant since θ_2 could be in the first or fourth quadrant.
Redundancy caused since θ_2 has two possible values

$$\theta_1 = \arcsin\left(\frac{l_2\sin(\theta_2)}{\sqrt{x^2+y^2}}\right) + \arctan 2\left(\frac{y}{x}\right)$$

The Algebraic Solution

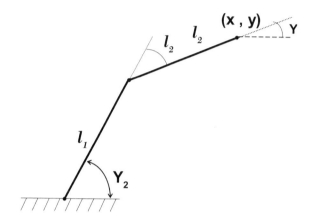

$c_1 = \cos\theta_1$

$c_{1+2} = \cos(\theta_2 + \theta_1)$

(1) $x = l_1 c_1 + l_2 c_{1+2}$

(2) $y = l_1 s_1 + l_2 \sin_{1+2}$

(3) $\theta = \theta_1 + \theta_2$

$(1)^2 + (2)^2 = x^2 + y^2 =$

$= (l_1^2 c_1^2 + l_2^2 (c_{1+2})^2 + 2l_1 l_2 c_1 (c_{1+2})) + (l_1^2 s_1^2 + l_2^2 (\sin_{1+2})^2 + 2l_1 l_2 s_1 (\sin_{1+2}))$

$= l_1^2 + l_2^2 + 2l_1 l_2 (c_1(c_{1+2}) + s_1(\sin_{1+2}))$

$= l_1^2 + l_2^2 + 2l_1 l_2 c_2 \leftarrow$ Only Unknown

$\therefore \theta_2 = \arccos\left(\frac{x^2 + y^2 - l_1^2 - l_2^2}{2l_1 l_2}\right)$

$x = l_1 c_1 + l_2 c_{1+2}$

$\quad = l_1 c_1 + l_2 c_1 c_2 - l_2 s_1 s_2$

$\quad = c_1(l_1 + l_2 c_2) - s_1(l_2 s_2)$

Note:

$\cos(a \overset{+}{_-} b) = (\cos a)(\cos b)\overset{-}{_+}(\sin a)(\sin b)$

$\sin(a \overset{-}{_-} b) = (\cos a)(\sin b)\overset{+}{_-}(\cos b)(\sin a)$

Note:

$\cos(a \overset{+}{_-} b) = (\cos a)(\cos b)\overset{-}{_+}(\sin a)(\sin b)$

$\sin(a \overset{+}{_-} b) = (\cos a)(\sin b)\overset{+}{_-}(\cos b)(\sin a)$

We know what θ_2 is from the previous slide. We need to solve for θ_1. Now we have two equations and two unknowns ($\sin \theta_1$ and $\sin \theta_1$).

$$x = l_1 c_1 + l_2 c_{1+2}$$
$$= l_1 c_1 + l_2 c_1 c_2 - l_2 s_1 s_2$$
$$= c_1 (l_1 + l_2 c_2) - s_1 (l_2 s_2)$$

Substituting for c1 and simplifying many times

$$c_1 = \frac{x + s_1 (l_2 s_2)}{(l_1 + l_2 c_2)}$$

$$y = \frac{x + s_1 (l_2 s_2)}{l_1 + l_2 c_2} (l_2 s_2) + s_1 (l_1 + l_2 c_2)$$

$$= \frac{1}{(l_1 + l_2 c_2)} (x l_2 s_2 + s_1 (l_1^2 + 2 l_1 l_2 c_2))$$

Notice this is the law of cosines and can be replaced by $x^2 + y^2$.

$$s_1 = \frac{y(l_1 + l_2 c_2) - x l_2 s_2}{x^2 + y^2}$$
$$\theta_1 = \arcsin \left(\frac{y(l_1 + l_2 c_2) - x l_2 s_2}{x^2 + y^2} \right)$$

5.4.5 Solution by a Systematic Approach

The above example is a simple two-link case, where the solution can be found out directly, since it contains only a few equations.

The elements of the left-hand-side matrix of the equation (section 5.4.3) are functions of n joints displacement variables. The elements of the right-hand-side matrix T are the desired position and orientation of the end-effector and are either constant or zero.

As the matrix equality implies element-by-element equality, 12 equations are obtained. To find the solution for n joint displacement variables from these 12 equations, the following guidelines are helpful.

(a) Look at the equations involving only one joint variable. Solve these equations first to get the corresponding joint variable solutions.

(b) Next, look for pairs or a set of equations that could be reduced to one equation in one joint variable by application of algebraic and trigonometry identities.

(c) Use the arc tangent (Atnt2) function instead of the arc cosine or arc sine functions. The two-argument Atnt2 (y, x) function returns the accurate angle in the range of such that θ lies between $-\pi$ to π. By examining the sign of both y and x and detecting whenever either x or y is 0.

(d) Solutions in terms of the element of the position vector components of ^0T_n are more efficient than those in terms of elements of the rotation matrix, as that latter may involve solving more complex equations.

(e) In the inverse kinematic model, the right-hand side of the equation (section 5.4.3) is known, while the left-hand side has n unknowns (q_1 , q_2......q_n). The left-hand side consists of products of n link transformation matrices, that is

$$^0T_n = {}^0T_1 \, {}^1T_2 \, {}^2T_{3....} \, {}^{n-1}T_n = T.$$

Recall that each i-1Ti is a function of only one unknown qi . Premultiplying both sides by the inverse of 0T1 yields

$$^1T_n = {}^1T_2 \, {}^2T_{3....} \, {}^{n-1}T_n = [{}^0T_1]^{-1} T.$$

The left-hand side of the equation now has (n-1) unknowns (q_2, q_3, ... , q_n) and the right-hand side matrix has only one unknown, the q_1. The matrix elements of the right-hand side are zero, constant, or the function of the joint variable q_1. A new set of 12 equations is obtained and it may now be possible to determine q_1 from the results of equations using guideline (a) or (b) above. Similarly, by postmultiplying both sides of the equation by the inverse of $^{n-1}T_n$, unknown equation q_n can be determined. This process can be repeated by solving for one unknown at a time, sequentially from q_1 to q_n or q_n to q_1, until all like unknown are found. This is known as the inverse transform approach.

The above systematic approach can be applied to find the inverse kinematic solution of manipulators having more than 2 DOF. The interested reader can refer to texts of robotic manipulators for examples of inverse kinematics involving more degrees of freedom.

6 CLASSIFICATION OF SENSORS

6.1 CLASSIFICATION OF SENSORS

There are a wide variety of sensors used in mobile robots (Figure 6.1). Some sensors are used to measure simple values like the internal temperature of a robot's electronics or the rotational speed of the motors. Other, more sophisticated sensors can be used to acquire information about the robot's environment or even to directly measure a robot's global position. We classify sensors using two important functional axes:

■ proprioceptive/exteroceptive and
■ passive/active.

241

(a)

(b)

FIGURE 6.1   **Examples of robots with multisensor systems. (a) Helpmate from Transition Research Corporation; (b) BIBA Robot, BlueBotics SA.**

Proprioceptive sensors measure values internal to the system (robot); for example, motor speed, wheel load, robot arm joint angles, and battery voltage.

Exteroceptive sensors acquire information from the robot's environment; for example, distance measurements, light intensity, and sound amplitude. Hence, exteroceptive sensor measurements are interpreted by the robot in order to extract meaningful environmental features.

Passive sensors measure ambient environment energy entering the sensor. Examples of passive sensors include temperature probes, microphones, and CCD or CMOS cameras.

Active sensors emit *energy into the environment*, and then measure the environmental reaction. Because active sensors can manage more controlled interactions with the environment, they often achieve superior performance. However, active sensing includes several risks: the outbound energy may affect the very characteristics that the sensor is attempting to measure. Furthermore, an active sensor may suffer from interference between its signal and those beyond its control. For example, signals emitted by other nearby robots, or similar sensors on the same robot may influence the resulting measurements. Examples of active sensors include wheel quadrature encoders, ultrasonic sensors, and laser range finders.

The table below gives a classification of the most useful sensors for mobile robot applications.

TABLE 6.1   Classification of Sensors Used in Mobile Robotics Applications

General Classification (typical use)	Sensor (Sensor System)	PC or EC	A or P
Tactile Sensors (detection of physical contact or closeness; security switches) Wheel/motor sensors (wheel/motor speed and position)	Contact switches, Bumpers,	EC	P
	Optical barriers,	EC	A
	Noncontact proximity sensors	EC	A
	Brush encoders	PC	P
	Potentiometers	PC	P
	Synchroes, Resolvers	PC	A
	Optical encoders	PC	A
	Magnetic encoders	PC	A
	Inductive encoders	PC	A
Heading sensors (orientation of the robot in relation to a fixed reference frame) Ground-based beacons (localization in a fixed reference frame)	Capacitive encoders	PC	A
	Compass	EC	P
	Gyroscope	PC	P
	Inclinometers	EC	A/P
	GPS	EC	A
	Active optical or RF beacons	EC	A
Active ranging (reflectivity, time-of-flight, and geometric triangulation) Laser rangefinder	Active ultrasonic beacons	EC	A
	Reflective beacons	EC	A
	Reflective sensors	EC	A
	Ultrasonic sensors	EC	A
	EC	A	A
Motion/speed sensors (speed relative to fixed or moving objects) Vision-based sensors (visual ranging, whole-image analysis, segmentation, object recognition)	Optical triangulation (1D)	EC	AAAAA
	Structured light (2D)	EC	A
	Doppler radar	EC	A
	CCD/CMOS camera(s)	EC	A
	Visual ranging packages	EC	A
	Object tracking packages		

Where:
    A: Active
    P: Passive
    A/P: Active/Passive
    PC: Proprioceptive
    EC: Exteroceptive

The sensor classes in Table 6.1 are arranged in ascending order of complexity and descending order of technological maturity. Tactile sensors and prospective sensors are critical to virtually all mobile robots, and are well understood and easily implemented. Commercial quadrature encoders, for example, may be purchased as part of a gear-motor assembly used in a mobile robot. At the other extreme, visual interpretation by means of one or more CCD/CMOS cameras provides a broad array of potential functionalities, from obstacle avoidance and localization to human face recognition. However, commercially available sensor units that provide visual functionalities are only now beginning to emerge.

6.2   ENCODERS AND DEAD RECKONING

Dead reckoning (derived from "deduced reckoning" of sailing days) is a simple mathematical procedure for determining the present location of a vessel by advancing some previous position through known course and velocity information over a given length of time. The vast majority of land-based mobile robotic systems in use today rely on dead reckoning to form the very backbone of their navigation strategy, and like their nautical counterparts, periodically null out accumulated errors with recurring "fixes" from assorted navigation aids.

The most simplistic implementation of dead reckoning is sometimes termed odometry; the term implies vehicle displacement along the path of travel is directly derived from some onboard "odometer." A common means of odometry instrumentation involves optical encoders directly coupled to the motor armatures or wheel axles.

Since most mobile robots rely on some variation of wheeled locomotion, a basic understanding of sensors that accurately quantify angular position and velocity is an important prerequisite to further discussions of odometry. There are a number of different types of rotational displacement and velocity sensors in use today:

- Brush encoders.
- Potentiometers.
- Synchros.
- Resolvers.
- Optical encoders.
- Magnetic encoders.
- Inductive encoders.
- Capacitive encoders.

A multitude of issues must be considered in choosing the appropriate device for a particular application. For mobile robot applications, incremental and absolute optical encoders are the most popular type. We will discuss those in the following sections.

Optical Encoders

The first optical encoders were developed in the mid-1940s by the Baldwin Piano Company for use as "tone wheels" that allowed electric organs to mimic other musical instruments. Today's corresponding devices basically embody a miniaturized version of the break-beam proximity sensor. A focused beam of light aimed at a matched photo detector is periodically interrupted by a coded opaque/transparent pattern on a rotating intermediate disk attached to the shaft of interest. The rotating disk may take the form of chrome on glass, etched metal, or photoplast such as mylar. Relative to the more complex alternating-current resolvers, the straightforward encoding scheme and inherently digital output of the optical encoder result in a low-cost reliable package with good noise immunity.

There are two basic types of optical encoders: incremental and absolute. The incremental version measures rotational velocity and can infer relative position, while absolute models directly measure angular position and infer velocity. If nonvolatile position information is not a consideration, incremental encoders generally are easier to interface and provide equivalent resolution at a much lower cost than absolute optical encoders.

Incremental Optical Encoders

The simplest type of incremental encoder is a single-channel tachometer encoder, basically an instrumented mechanical light chopper that produces a certain number of sine- or square-wave pulses for each shaft revolution. Adding pulses increases the resolution (and subsequently the cost) of the unit. These relatively inexpensive devices are well suited as velocity feedback sensors in medium- to high-speed control systems, but run into noise and stability problems at extremely slow velocities due to quantization errors. The tradeoff here is resolution versus update rate: improved transient response requires a faster update rate, which for a given line count reduces the number of possible encoder pulses per sampling interval.

In addition to low-speed instabilities, single-channel tachometer encoders are also incapable of detecting the direction of rotation and thus cannot be used as position sensors. Phase-quadrature incremental encoders overcome these problems by adding a second channel, displaced from the first, so the resulting pulse trains are 90 degrees out of phase as shown in Figure 6.2. This technique allows the decoding electronics to determine which channel is leading the other and hence ascertain the direction of rotation, with the added benefit of increased resolution.

The incremental nature of the phase-quadrature output signals dictates that any resolution of angular position can only be relative to some specific reference, as opposed to absolute. Establishing such a reference can be accomplished in a number of ways. For applications involving continuous 360-degree rotation, most encoders incorporate as a third channel a special index output that goes high once for each complete revolution of the shaft (see Figure 6.2).

Intermediate shaft positions are then specified by the number of encoder up counts or down counts from this known index position. One disadvantage of this

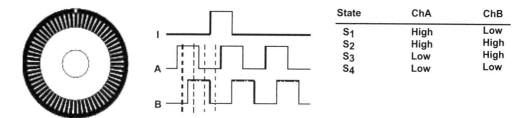

State	ChA	ChB
S₁	High	Low
S₂	High	High
S₃	Low	High
S₄	Low	Low

FIGURE 6.2 The observed phase relationship between Channel A and B pulse trains can be used to determine the direction of rotation with a phase-quadrature encoder, while unique output states S - S allow for up to a four-fold increase in resolution. The single slot in the outer track generates one index pulse per disk rotation.

approach is that all relative position information is lost in the event of a power interruption.

In the case of limited rotation, such as the back-and-forth motion of a pan or tilt axis, electrical limit switches and/or mechanical stops can be used to establish a home reference position. To improve repeatability this homing action is sometimes broken into two steps. The axis is rotated at reduced speed in the appropriate direction until the stop mechanism is encountered, whereupon rotation is reversed for a short predefined interval. The shaft is then rotated slowly back into the stop at a specified low velocity from this designated start point, thus eliminating any variations in inertial loading that could influence the final homing position. This two-step approach can usually be observed in the power-on initialization of stepper-motor positioners for dot-matrix printer heads.

Alternatively, the absolute indexing function can be based on some external referencing action that is decoupled from the immediate servo-control loop. A good illustration of this situation involves an incremental encoder used to keep track of platform steering angles. For example, when the K2A Navmaster [CYBERMOTION] robot is first powered up, the absolute steering angle is unknown, and must be initialized through a "referencing" action with the docking beacon, a nearby wall, or some other identifiable set of landmarks of known orientation. The up/down count output from the decoder electronics is then used to modify the vehicle heading register in a relative fashion.

A growing number of very inexpensive off-the-shelf components have contributed to making the phase-quadrature incremental encoder the rotational sensor of choice within the robotics research and development community. Several manufacturers now offer small DC gear-motors with incremental encoders already attached to the armature shafts. Within the U.S. automated guided vehicle (AGV) industry, however, resolves are still generally preferred over optical encoders for their perceived superiority under harsh operating conditions.

Interfacing an incremental encoder to a computer is not a trivial task. A simple state-based interface as implied in Figure 6.2 is inaccurate if the encoder changes direction at certain positions and false pulses can result from the interpretation of the sequence of state changes.

A more versatile encoder interface is the HCTL 1100 motion controller chip made by Hewlett Packard [HP]. The HCTL chip performs not only accurate quadrature decoding of the incremental wheel encoder output, but it provides many important additional functions, including among others:

- closed-loop position control,
- closed-loop velocity control in P or PI fashion,
- 24-bit position monitoring.

The HCTL 1100 has been tested and used in many different mobile robot control interfaces. The chip has proven to work reliably and accurately, and it is used on commercially available mobile robots, such as the TRC LabMate and HelpMate. The HCTL 1100 costs only $40 and it comes highly recommended.

Absolute Optical Encoders

Absolute encoders are typically used for slower rotational applications that require positional information when potential loss of reference from power interruption cannot be tolerated. Discrete detector elements in a photovoltaic array are individually aligned in break-beam fashion with concentric encoder tracks as shown in Figure 6.3, creating, in effect, a noncontact implementation of a commutating

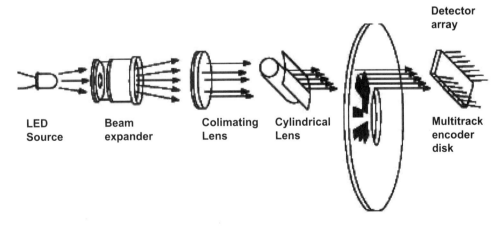

LED   **Beam**   **Colimating**   **Cylindrical**   **Detector array**
Source   **expander**   **Lens**   **Lens**   **Multitrack encoder disk**

FIGURE 6.3 A line source of light passing through a coded pattern of opaque and transparent segments on the rotating encoder disk results in a parallel output that uniquely specifies the absolute angular position of the shaft (adapted from [Agent, 1991]).

brush encoder. The assignment of a dedicated track for each bit of resolution results in larger-size disks (relative to incremental designs), with a corresponding decrease in shock and vibration tolerance. A general rule of thumb is that each additional encoder track doubles the resolution but quadruples the cost.

Instead of the serial bit streams of incremental designs, absolute optical encoders provide a parallel word output with a unique code pattern for each quantized shaft position. The most common coding schemes are Gray code, natural binary and binary-coded decimal. The Gray code is characterized by the fact that only one bit changes at a time, a decided advantage in eliminating asynchronous ambiguities caused by electronic and mechanical component tolerances (see Figure 6.4a). Binary code, on the other hand, routinely involves multiple bit changes when incrementing or decrementing the count by one. For example, when going from position 255 to position 0 in Figure 6.4b, eight bits toggle from 1s to 0s. Since there is no guarantee all threshold detectors monitoring the detector elements tracking each bit will toggle at the same precise instant, considerable ambiguity can exist during state transition with a coding scheme of this form. Some type of handshake line signaling valid data available would be required if more than one bit were allowed to change between consecutive encoder positions.

Absolute encoders are best suited for slow and/or infrequent rotations such as steering angle encoding, as opposed to measuring high-speed continuous (i.e., drive wheel) rotations as would be required for calculating displacement along the path of travel. Although not quite as robust as resolvers for high-temperature, high-shock applications, absolute encoders can operate at temperatures over 125 °C, and medium-resolution (1,000 counts per revolution) metal or mylar disk

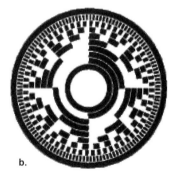

a.   b.

FIGURE 6.4   Rotating an 8-bit absolute Gray-code disk. a. Counterclockwise rotation by one position increment will cause only one bit to change. b. The same rotation of a binary-coded disk will cause all bits to change in this particular case (255 to 0) illustrated by the reference line at twelve o'clock.

designs can compete favorably with resolvers in terms of shock resistance. A potential disadvantage of absolute encoders is their parallel data output, which requires a more complex interface due to the large number of electrical leads. A 13-bit absolute encoder using complimentary output signals for noise immunity would require a 28-conductor cable (13 signal pairs plus power and ground), versus only six for a resolver or incremental encoder.

6.3   INFRARED SENSORS

Theory of Operation

A line sensor in its simplest form is a sensor capable of detecting a contrast between adjacent surfaces, such as difference in color, roughness, or magnetic properties. The simplest would be detecting a difference in color, for example, black and white surfaces. Using simple optoelectronics, such as infrared phototransistors, color contrast can easily be detected. Infrared emitter/detectors or phototransistors are inexpensive and are easy to interface to a microcontroller.

The theory of operation is simple and for brevity, only the basics will be considered. For more information about the physics of these sensors, please refer to an optoelectronics and heat transfer text. For now we will consider the basic effects of light and what happens when it shines on a black or white surface. When light shines on a white surface, most of the incoming light is reflected away from the surface. In contrast, most of the incoming light is absorbed if the surface is black. Therefore, by shining light on a surface and having a sensor to detect the amount of light that is reflected, a contrast between black and white surfaces can be detected. Figure 6.5 shows an illustration of the basics just covered.

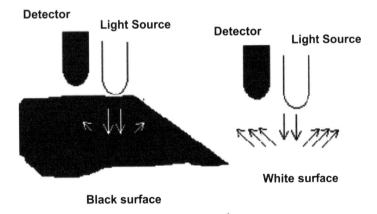

FIGURE 6.5   Light reflecting off a white and black surface. More light is reflected from the white surface compared to the black surface.

Using what we know about black and white surfaces, the objective of tracking a line is simple and can be achieved using the appropriate sensors. In this section, we will consider the use of two pairs of emitters and detectors. The drive configuration for the robot is assumed to be differential, i.e., like the tracks of an army tank vehicle. Two pairs of sensors are used to keep the robot on the line as it moves. Each sensor output is monitored to determine the location of the tape relative to the robot. The main objective of the robot is to position itself such that the tape line falls between the two extreme sensors. If the tape line ever ventures past these two extreme sensors, then the robot corrects by turning in the appropriate direction to maintain tracking. Two different types of light sensors set up in the configuration will be used for line tracking.

6.4   GROUND-BASED RF SYSTEMS

Ground-based RF position location systems are typically of two types:

- Passive hyperbolic line-of-position phase-measurement systems that compare the time-of-arrival phase differences of incoming signals simultaneously emitted from surveyed transmitter sites.
- Active radar-like trilateration systems that measure the round-trip propagation delays for a number of fixed-reference transponders. Passive systems are generally preferable when a large number of vehicles must operate in the same local area, for obvious reasons.

6.4.1   LORAN

An early example of the first category is seen in LORAN (short for long range navigation). Developed at MIT during World War II, such systems compare the time of arrival of two identical signals broadcast simultaneously from high-power transmitters located at surveyed sites with a known separation baseline. For each finite time difference (as measured by the receiver) there is an associated hyperbolic line of position as shown in Figure 6.6. Two or more pairs of master/slave stations are required to get intersecting hyperbolic lines resulting in a two-dimensional (latitude and longitude) fix.

The original implementation (LORAN A) was aimed at assisting convoys of liberty ships crossing the North Atlantic in stormy winter weather. Two 100 kW slave transmitters were located about 200 miles on either side of the master station. Non-line-of-sight ground-wave propagation at around 2 MHz was employed, with pulsed as opposed to continuous-wave transmissions to aid in sky-wave discrimination. The time-of-arrival difference was simply measured as the lateral separation of the two pulses on an oscilloscope display, with a typical

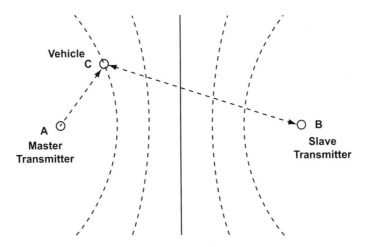

FIGURE 6.6   **For each hyperbolic line-of-position, length ABC minus length AC equals some constant K.**

accuracy of around 1 μs. This numerical value was matched to the appropriate line of position on a special LORAN chart of the region, and the procedure then repeated for another set of transmitters. For discrimination purposes, four different frequencies were used, 50 kHz apart, with 24 different pulse repetition rates in the neighborhood of 20 to 35 pulses per second. In situations where the hyperbolic lines intersected more or less at right angles, the resulting (best-case) accuracy was about 1.5 kilometers.

LORAN A was phased out in the early '80s in favor of LORAN C, which achieves much longer over-the-horizon ranges through use of 5 MW pulses radiated from 400-meter (1,300 ft.) towers at a lower carrier frequency of 100 kHz. For improved accuracy, the phase differences of the first three cycles of the master and slave pulses are tracked by phase-lock-loops in the receiver and converted to a digital readout, which is again cross-referenced to a preprinted chart. Effective operational range is about 1,000 miles, with best-case accuracies in the neighborhood of 100 meters (330 ft.). Coverage is provided by about 50 transmitter sites to all U.S. coastal waters and parts of the North Atlantic, North Pacific, and the Mediterranean.

6.4.2   Kaman Sciences Radio Frequency Navigation Grid

The Unmanned Vehicle Control Systems Group of Kaman Sciences Corporation, Colorado Springs, CO, has developed a scaled-down version of a LORAN-type hyperbolic position-location system known as the Radio Frequency Navigation Grid (RFNG). The original application in the late 1970s involved

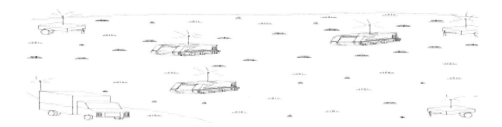

FIGURE 6.7 Kaman sciences 1500 W navigation grid is a scaled-down version of the LORAN concept, covering an area 8 to 15 km on a side with a position location repeatability of 1 m.

autonomous route control of unmanned mobile targets used in live-fire testing of the laser-guided Copperhead artillery round. The various remote vehicles sense their position by measuring the phase differences in received signals from a master transmitter and two slaves situated at surveyed sites within a 30 km² (18.75 mi²) area as shown in Figure 6.7. System resolution is 3 centimeters (1.5 in.) at a 20 Hz update rate, resulting in a vehicle positioning repeatability of 1 meter (3.3 ft.).

Path trajectories are initially taught by driving a vehicle over the desired route and recording the actual phase differences observed. This file is then played back at run time and compared to measured phase difference values, with vehicle steering servoed in an appropriate manner to null any observed error signal. Velocity of advance is directly controlled by the speed of file playback. Vehicle speeds in excess of 50 km/h (30 mph) are supported over path lengths of up to 15 kilometers (9.4 mi.). Multiple canned paths can be stored and changed remotely, but vehicle travel must always begin from a known start point due to an inherent 6.3 meters (20 ft.) phase ambiguity interval associated with the grid.

The Threat Array Control and Tracking Information Center (TACTIC) is offered by Kaman Sciences to augment the RFNG by tracking and displaying the location and orientation of up to 24 remote vehicles. Real-time telemetry and recording of vehicle heading, position, velocity, status, and other designated parameters (i.e., fuel level, oil pressure, and battery voltage) are supported at a 1 Hz update rate. The TACTIC operator has direct control over engine start, automatic path playback, vehicle pause/resume, and emergency halt functions. Non-line-of-sight operation is supported through use of a 23.825 MHz grid frequency in conjunction with a 72 MHz control and communications channel.

6.4.3 Precision Location Tracking and Telemetry System

Precision Technology, Inc., of Saline, MI, has recently introduced to the automotive racing world an interesting variation of the conventional phase-shift measurement approach (type 1 RF system). The company's precision location tracking and telemetry system employs a number of receive-only antennae situated at fixed locations around a racetrack to monitor a continuous sine wave transmission from a moving vehicle. By comparing the signals received by the various antennae to a common reference signal of identical frequency generated at the base station, relative changes in vehicle position with respect to each antenna can be inferred from resulting shifts in the respective phase relationships. The 58 MHz VHF signal allows for non-line-of-sight operation, with a resulting precision of approximately 1 to 10 centimeters (0.4 to 4 in.). From a robotics perspective, problems with this approach arise when more than one vehicle must be tracked. The system costs $200,000 to $400,000, depending on the number of receivers used, but the system is not suitable for indoor operations.

6.4.4 Motorola Mini-ranger Falcon

An example of the active transponder category of ground-based RF position-location techniques is seen in the Mini-ranger Falcon series of range positioning systems offered by the Government and Systems Technology Group of Motorola. The Falcon 484 configuration depicted in Figure 6.8 is capable of measuring line-of-sight distances from 100 meters (328 ft.) out to 75 kilometers (47 miles). An initial calibration is performed at a known location to determine the turnaround delay (TAD) for each transponder (i.e., the time required to transmit a response back to the interrogator after receipt of interrogation). The actual distance between the interrogator and a given transponder is found by:

$$D = \frac{(T_c - T_d)C}{2}.$$

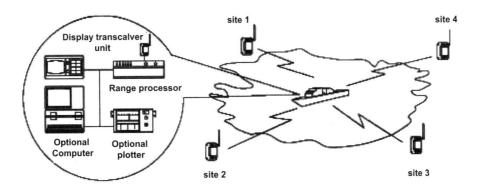

FIGURE 6.8 Motorola's Mini-ranger Falcon 484 R position-location system provides 2 m (6.5 ft.) accuracy over ranges of 100 m to 75 km (328 ft. to 47 mi.).

Where,

D = separation distance
Te = total elapsed time
Td = transponder turnaround delay
c = speed of light.

The MC6809-based range processor performs a least-squares position solution at a 1 Hz update rate, using range inputs from two, three, four, or 16 possible reference transponders. The individual reference stations answer only to uniquely coded interrogations and operate in C-band (5410 to 5890 MHz) to avoid interference from popular X-band marine radars. Up to 20 mobile users can time share the Falcon 484 system (50 ms per user maximum). System resolution is in tenths of units (m., ft., or yd.) with a range accuracy of 2 meters (6.5 ft.) probable.

Power requirements for the fixed-location reference stations are 22 to 32 VDC at 13 W nominal, 8.5 W standby, while the mobile range processor and its associated transmitter-receiver and display unit draw 150 W at 22 to 32 VDC. The Falcon system comes in different, customized configurations. Complete system cost is $75,000 to $100,000.

6.4.5 Harris Infogeometric System

Harris Technologies, Inc., is developing a ground-based R-position location and communications strategy wherein moderately priced infogeometric (IG) devices cooperatively form self-organizing instrumentation and communication networks. Each IG device in the network has full awareness of the identity, location, and orientation of all other IG devices and can communicate with other such devices in both party-line and point-to-point communication modes.

The IG devices employ digital code-division-multiple-access (CDMA) spread-spectrum R hardware that provides the following functional capabilities:

- Network-level mutual autocalibration.
- Associative location and orientation tracking.
- Party-line and point-to-point data communications (with video and audio options).
- Distributed sensor data fusion.

Precision position location on the move is based on high-speed range trilateration from fixed reference devices, a method commonly employed in many instrumentation test ranges and other tracking system applications. In this approach, each beacon has an extremely accurate internal clock that is carefully synchronized with all other beacon clocks. A time-stamped (coded) R signal is periodically sent

TABLE 6.2 Raw Data Measurement Resolution and Accuracy		
Parameter	Resolution	Biasing
Range	1	5 m
	3.3	16.4 ft.
Bearing (Az, El)	2	20
Orientation (Az)	2	20

by each transmitter. The receiver is also equipped with a precision clock, so that it can compare the timing information and time of arrival of the incoming signals to its internal clock. This way, the system is able to accurately measure the signals' time of flight and thus the distance between the receiver and the three beacons. This method, known as "differential location regression" is essentially the same as the locating method used in global positioning systems (GPS).

To improve accuracy over current range-lateration schemes, the HTI system incorporates mutual data communications, permitting each mobile user access to the time-tagged range measurements made by fixed reference devices and all other mobile users. This additional network-level range and timing information permits more accurate time synchronization among device clocks, and automatic detection and compensation for uncalibrated hardware delays.

Each omnidirectional CDMA spread-spectrum "geometric" transmission uniquely identifies the identity, location, and orientation of the transmitting source. Typically the available geometric measurement update rate is in excess of 1,000 kHz. Harris quotes a detection radius of 500 meters (1,640 ft.) with 100 mW peak power transmitters. Larger ranges can be achieved with stronger transmitters. Harris also reports on "centimeter-class repeatability accuracy" obtained with a modified transmitter called an "interactive beacon." Tracking and communications at operating ranges of up to 20 kilometers (12.5 mi.) are also supported by higher transmission power levels of 1 to 3 W. Typical "raw data" measurement resolution and accuracies are cited in Table 6.2.

Enhanced tracking accuracies for selected applications can be provided as cited in Table 6.3. This significant improvement in performance is provided by sensor data

TABLE 6.3 Enhanced Tracking Resolution and Accuracies Obtained through Sensor Data Fusion		
Parameter	Resolution	Biasing
Range	0.1–0.3	0.1–0.3 m
	0.3–0.9	0.3–0.9 ft.
Bearing (Az, El)	0.5–1.0	0.5–1.0″
Orientation (Az)	0.5–1.0	0.5–1.0″

fusion algorithms that exploit the high degree of relational redundancy that is characteristic for infogeometric network measurements and communications.

Infogeometric enhancement algorithms also provide the following capabilities:

- Enhanced tracking in multipath and clutter—permits precision robotics tracking even when operating indoors.
- Enhanced near/far interference reduction—permits shared-spectrum operations in potentially large user networks (i.e., hundreds to thousands).

Operationally, mobile IG networks support precision tracking, communications, and command and control among a wide variety of potential user devices. A complete infogeometric positioning system is commercially available at a cost of $30,000 or more (depending on the number of transmitters required). The system also requires an almost clear "line of sight" between the transmitters and receivers. In indoor applications, the existence of walls or columns obstructing the path will dramatically reduce the detection range and may result in erroneous measurements, due to multipath reflections.

6.5   ACTIVE BEACONS

Active beacon navigation systems are the most common navigation aids on ships and airplanes. Active beacons can be detected reliably and provide very accurate positioning information with minimal processing. As a result, this approach allows high sampling rates and yields high reliability, but it does also incur high cost in installation and maintenance. Accurate mounting of beacons is required for accurate positioning. For example, land surveyors' instruments are frequently used to install beacons in a high-accuracy application. Kleeman notes that:

"Although special beacons are at odds with notions of complete robot autonomy in an unstructured environment, they offer advantages of accuracy, simplicity, and speed—factors of interest in industrial and office applications, where the environment can be partially structured."

One can distinguish between two different types of active beacon systems: trilateration and triangulation.

6.5.1   Trilateration

Trilateration is the determination of a vehicle's position based on distance measurements to known beacon sources. In trilateration navigation systems there are usually three or more transmitters mounted at known locations in the environment and one receiver on board the robot. Conversely, there may be one transmitter on board and the receivers are mounted on the walls. Using time-of-flight information, the system computes the distance between the stationary transmitters and the onboard receiver. Beacon systems based on ultrasonic sensors are another example.

6.5.2 Triangulation

In this configuration there are three or more active transmitters (usually infrared) mounted at known locations in the environment, as shown in Figure 6.9. A rotating sensor on board the robot registers the angles λ_1, λ_2, and λ_3 at which it "sees" the transmitter beacons relative to the vehicle's longitudinal axis. From these three measurements the unknown x- and y-coordinates and the unknown vehicle orientation θ can be computed. Simple navigation systems of this kind can be built very inexpensively. One problem with this configuration is that the active beacons need to be extremely powerful to insure omnidirectional transmission over large distances. Since such powerful beacons are not very practical it is necessary to focus the beacon within a cone-shaped propagation pattern. As a result, beacons are not visible in many areas, a problem that is particularly grave because at least three beacons must be visible for triangulation. A commercially available sensor system based on this configuration was tested at the University of Michigan in 1990. The system provided an accuracy of approximately ±5 centimeters (±2 in.), but the aforementioned limits on the area of application made the system unsuitable for precise navigation in large open areas.

Triangulation methods can further be distinguished by the specifics of their implementation:

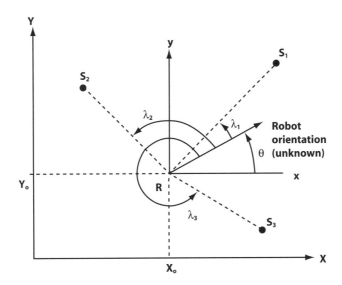

FIGURE 6.9   The basic triangulation problem: a rotating sensor head measures the three angles λ_1, λ_2, and λ_3 between the vehicle's longitudinal axes and the three sources s_1, s_2, and s_3.

a. **Rotating Transmitter-receiver, Stationary Reflectors:** In this implementation there is one rotating laser beam on board the vehicle and three or more stationary retroreflectors are mounted at known locations in the environment.

b. **Rotating Transmitter, Stationary Receivers:** Here the transmitter, usually a rotating laser beam, is used on board the vehicle. Three or more stationary receivers are mounted on the walls. The receivers register the incident beam, which may also carry the encoded azimuth of the transmitter.

For either one of the above methods, we will refer to the stationary devices as "beacons," even though they may physically be receivers, retroreflectors, or transponders.

6.5.3 Discussion on Triangulation Methods

In general, it can be shown that triangulation is sensitive to small angular errors when either the observed angles are small, or when the observation point is on or near a circle that contains the three beacons. Assuming reasonable angular measurement tolerances, it was found that accurate navigation is possible throughout a large area, although error sensitivity is a function of the point of observation and the beacon arrangements.

Three-point Triangulation

Cohen and Koss in 1992 performed a detailed analysis on three-point triangulation algorithms and ran computer simulations to verify the performance of different algorithms. The results are summarized as follows:

- The geometric triangulation method works consistently only when the robot is within the triangle formed by the three beacons. There are areas outside the beacon triangle where the geometric approach works, but these areas are difficult to determine and are highly dependent on how the angles are defined.
- The geometric circle intersection method has large errors when the three beacons and the robot all lie on, or close to, the same circle.
- The Newton-Raphson method fails when the initial guess of the robot's position and orientation is beyond a certain bound.
- The heading of at least two of the beacons was required to be greater than 90 degrees. The angular separation between any pair of beacons was required to be greater than 45 degrees.

In summary, it appears that none of the above methods alone is always suitable, but an intelligent combination of two or more methods helps overcome the individual weaknesses.

Yet another variation of the triangulation method is the so-called running fix. The underlying principle of the running fix is that an angle or range obtained from a beacon at time t-1 can be utilized at time t, as long as the cumulative movement vector recorded since the reading was obtained is added to the position vector of the beacon, thus creating a virtual beacon.

6.5.4   Triangulation with More than Three Landmarks

An algorithm, called the position estimator, is used to solve the general triangulation problem. This problem is defined as follows: given the global position of n landmarks and corresponding angle measurements, estimate the position of the robot in the global coordinate system. The n landmarks are represented as complex numbers and the problem is formulated as a set of linear equations. By contrast, the traditional law-of-cosines approach yields a set of nonlinear equations. The algorithm only fails when all landmarks are on a circle or a straight line. The algorithm estimates the robot's position in O(n) operations where n is the number of landmarks on a two-dimensional map.

Compared to other triangulation methods, the position estimator algorithm has the following advantages:

(1) The problem of determining the robot position in a noisy environment is linearized,
(2) The algorithm runs in an amount of time that is a linear function of the number of landmarks,
(3) The algorithm provides a position estimate that is close to the actual robot position, and
(4) Large errors ("outliers") can be found and corrected.

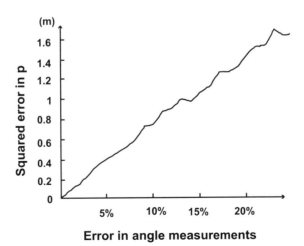

FIGURE 6.10   Simulations result using the position estimator algorithm on an input of noisy angle measurements. The squired error in the position estimate p (in meters) is shown as a function of measurement errors (in percent of the actual angle).

The results of a simulation for the following scenario are presented: the robot is at the origin of the map, and the landmarks are randomly distributed in a 10x10 meter (32x32 ft.) area.

The robot is at the corner of this area. The distance between a landmark and the robot is at most 14.1 meters (46 ft.) and the angles are at most 45 degrees. The simulation results show that large errors due to misidentified landmarks and erroneous angle measurements can be found and discarded. Subsequently, the algorithm can be repeated without the outliers, yielding improved results. One example is shown in Figure 6.11, which depicts simulation results using the algorithm position estimator. The algorithm works on an input of 20 landmarks (not shown in Figure 6.11) that were randomly placed in a 10×10 meter (32×32 ft.) workspace. The simulated robot is located at (0,0). Eighteen of the landmarks were simulated to have a one-percent error in the angle measurement and two of the landmarks were simulated to have a large 10-percent angle measurement error. With the angle measurements from 20 landmarks the position estimator produces 19 position estimates p1–p19 (shown as small blobs in Figure 6.11). Averaging these 19 estimates yields the computed robot position. Because of the two landmarks with large angle measurement errors, two position estimates are bad: p_5 at (79 cm, 72 cm) and p_{18} at (12.5 cm, 18.3 cm).

Because of these poor position estimates, the resulting centroid (average) is at P^a = (17 cm, 24 cm). However, the position estimator can identify and exclude the two outliers. The centroid calculated without the outliers p5 and p18 is at P^b = (12.5 cm, 18.3 cm). The final position estimate after the position estimator is applied again on the 18 "good" landmarks (i.e., without the two outliers) is at P^c = (6.5 cm, 6.5 cm).

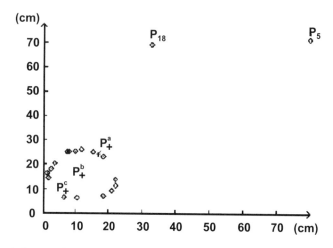

FIGURE 6.11   Simulation results showing the effect of outliers and the result of removing the outliers.

6.6   ULTRASONIC TRANSPONDER TRILATERATION

Ultrasonic trilateration schemes offer a medium- to high-accuracy, low-cost solution to the position location problem for mobile robots. Because of the relatively short range of ultrasound, these systems are suitable for operation in relatively small work areas and only if no significant obstructions are present to interfere with wave propagation. The advantages of a system of this type fall off rapidly, however, in large multiroom facilities due to the significant complexity associated with installing multiple networked beacons throughout the operating area.

Two general implementations exist: 1) a single transducer transmitting from the robot, with multiple fixed-location receivers, and 2) a single receiver listening on the robot, with multiple fixed transmitters serving as beacons. The first of these categories is probably better suited to applications involving only one or at most a very small number of robots, whereas the latter case is basically unaffected by the number of passive receiver platforms involved (i.e., somewhat analogous to the Navstar GPS concept).

6.6.1   IS Robotics 2D Location System

IS Robotics, Inc., Somerville, MA, a spin-off company from MIT's renowned Mobile Robotics Lab, has introduced a beacon system based on an inexpensive ultrasonic trilateration system. This system allows their Genghis series robots to

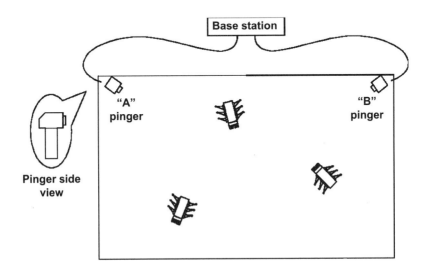

FIGURE 6.12   The ISR Genghis series of legged robots localize x-y position with a master/slave trilateration scheme using two 40 KHz ultrasonic "pingers".

localize position to within 12.7 millimeters (0.5 in.) over a 9.1×9.1 meter (30×30 ft.) operating area. The ISR system consists of a base station master hard-wired to two slave ultrasonic "pingers" positioned a known distance apart (typically 2.28 m — 90 in.) along the edge of the operating area as shown in Figure 6.12. Each robot is equipped with a receiving ultrasonic transducer situated beneath a cone-shaped reflector for omnidirectional coverage. Communication between the base station and individual robots is accomplished using a Proxim spread-spectrum (902 to 928 MHz) RF link.

The base station alternately fires the two 40-kHz ultrasonic pingers every half second, each time transmitting a two-byte radio packet in broadcast mode to advise all robots of pulse emission. Elapsed time between radio packet reception and detection of the ultrasonic wave front is used to calculate distance between the robot's current position and the known location of the active beacon. Inter robot communication is accomplished over the same spread-spectrum channel using a time-division multiple-access scheme controlled by the base station. Principle sources of error include variations in the speed of sound, the finite size of the ultrasonic transducers, nonrepetitive propagation delays in the electronics, and ambiguities associated with time-of-arrival detection. The cost for this system is $10,000.

6.6.2   Tulane University 3D Location System

Researchers at Tulane University in New Orleans, LA, have come up with some interesting methods for significantly improving the time-of-arrival measurement accuracy for ultrasonic transmitter-receiver configurations, as well as compensating for the varying effects of temperature and humidity. In the hybrid scheme illustrated in Figure 6.13, envelope peak detection is employed to establish the approximate time of signal arrival, and to consequently eliminate

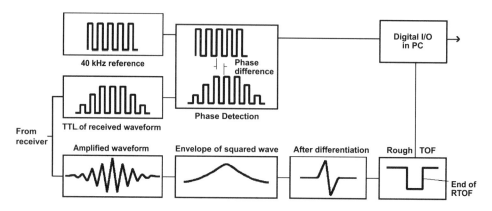

FIGURE 6.13   A combination of threshold adjusting and phase detection is employed to provide higher accuracy in time-of-arrival measurements in the Tulane University ultrasonic position-locator system.

ambiguity interval problems for a more precise phase-measurement technique that provides final resolution. The desired 0.025 millimeter (0.001 in.) range accuracy required a time unit discrimination of 75 nanoseconds at the receiver, which can easily be achieved using fairly simplistic phase measurement circuitry, but only within the interval of a single wavelength. The actual distance from transmitter to receiver is the summation of some integer number of wavelengths (determined by the coarse time-of-arrival measurement) plus that fractional portion of a wavelength represented by the phase measurement results.

The set of equations describing time-of-flight measurements for an ultrasonic pulse propagating from a mobile transmitter located at point (u, v, w) to various receivers fixed in the inertial reference frame can be listed in matrix form as follows:

$$
\begin{Bmatrix}
(t_1 - t_2) \\
(t_2 - t_d) \\
* \\
* \\
* \\
(t_n - t_d)
\end{Bmatrix}
=
\begin{bmatrix}
1 & r_1^2 & 2x_1 & 2y_1 & 2z_1 \\
1 & r_2^2 & 2x_2 & 2y_2 & 2z_2 \\
* & & & & \\
* & & & & \\
* & & & & \\
1 & r_n^2 & 2x_n & 2y_n & 2z_n
\end{bmatrix}
\begin{Bmatrix}
\dfrac{p^2}{c^2} \\
\dfrac{1}{c^2} \\
-\dfrac{u}{c^2} \\
-\dfrac{v}{c^2} \\
-\dfrac{w}{c^2}
\end{Bmatrix}
$$

whereas:

ti = measured time of flight for transmitted pulse to reach i^{th} receiver
td = system throughput delay constant
ri² = sum of squares of ith receiver coordinates
(x_i, y_i, z_i) = location coordinates of ith receiver
(u, v, w) = location coordinates of mobile transmitter
c = speed of sound
p^2 = sum of squares of transmitter coordinates.

The above equation can be solved for the vector on the right to yield an estimated solution for the speed of sound c, transmitter coordinates (u, v, w), and an independent term p^2 that can be compared to the sum of the squares of the transmitter coordinates as a checksum indicator. An important feature of this representation is the use of an additional receiver (and associated equation) to enable treatment of the speed of sound itself as an unknown, thus ensuring continuous on-the-fly recalibration to account for temperature and humidity effects.

(The system throughput delay constant t_d can also be determined automatically from a pair of equations for $1/c^2$ using two known transmitter positions. This procedure yields two equations with t_d and c as unknowns, assuming c remains constant during the procedure.) A minimum of five receivers is required for an unambiguous three-dimensional position solution, but more can be employed to achieve higher accuracy using a least-squares estimation approach. Care must be taken in the placement of receivers to avoid singularities.

Figueroa and Mahajan report a follow-up version intended for mobile robot positioning that achieves 0.25 millimeter (0.01 in.) accuracy with an update rate of 100 Hz. The prototype system tracks a TRC LabMate over a 2.7×3.7 meter (9×12 ft.) operating area with five ceiling-mounted receivers and can be extended to larger floor plans with the addition of more receiver sets. An RF link will be used to provide timing information to the receivers and to transmit the subsequent x-y position solution back to the robot. Three problem areas are being further investigated to increase the effective coverage and improve resolution:

- Actual transmission range does not match the advertised operating range for the ultrasonic transducers, probably due to a resonant frequency mismatch between the transducers and electronic circuitry.
- The resolution of the clocks (6 MHz) used to measure time of flight is insufficient for automatic compensation for variations in the speed of sound.
- The phase-detection range-measurement correction sometimes fails when there is more than one wavelength of uncertainty. This problem can likely be solved using the frequency division scheme described by Figueroa and Barbieri.

1490 Digital Compass Sensor

This sensor provides eight directions of heading information by measuring the earth's magnetic field using hall-effect technology. The 1490 sensor is internally designed to respond to directional change similar to a liquid-filled compass. It will return to the indicated direction from a 90-degree displacement in approximately 2.5 seconds with no overswing. The 1490 can operate tilted up to 12 degrees with acceptable error. It is easily interfaced to digital circuitry and microprocessors using only pull-up resistors.

Specifications

Power	5–18 volts DC @ 30 ma
Outputs	Open collector NPN, sink 25 ma per direction
Weight	2.25 grams
Size	12.7 mm diameter, 16 mm tall
Pins	3 pins on 4 sides on .050 centers
Temp	-20 to +85 degrees C

How to Add a Digital Compass to the PPRK

Overview (Palm Pilot Robot Kit)

A digital compass can be very useful for mobile robot navigation, especially for a small robot such as the PPRK, which lacks wheel encoders and hence built-in odometry and dead reckoning. Dinsmore Instrument Co. produces a very low-cost ($14) digital compass, the 1490, which can be easily interfaced to the SV203 board of the PPRK. The compass is shown in Figure 6.14:

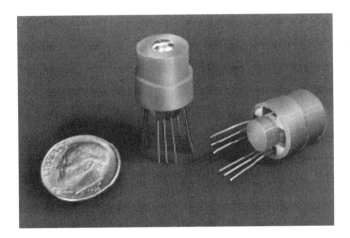

FIGURE 6.14

Interfacing

The compass provides eight headings (N, NE, E, SE, S, SW, W, and NW), which are encoded in four signal wires (N, E, S, W). Each of the wires is standard TTL open-collector NPN output and can be interfaced to digital input lines via pull-up resistors.

However, the SV203 has no digital input lines—instead, it has five analog voltage ports, three of which are already used by the IR sensors. It is still possible to interface the compass to the SV203 by converting the four digital signals into analog voltage and reading this voltage through a remaining analog port. The circuit below is based on a standard resistor-ladder digital-to-analog converter with four bits, with the addition of four pull-up resistors. Although these resistors lead to deviations of the converted voltage from exact powers of two, this circuit only has to encode eight different values for the possible headings, and the choice of resistors in the circuit results in clear separation between the voltages corresponding to different headings.

The transistors shown in the circuit are inside the compass—only the resistors have to be supplied. The compass has 12 pins:

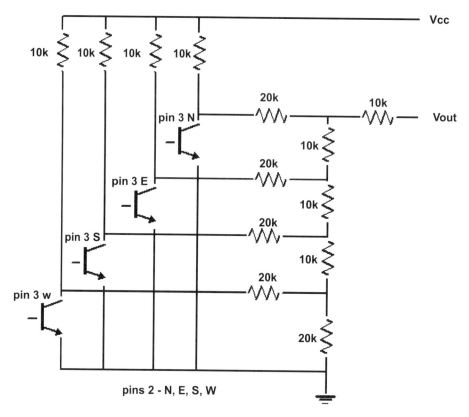

FIGURE 6.15

1N, 1E, 1S, 1W—Vcc, connect to pin 9 of SV203's port A (J3);
2N, 2E, 2S, 2W—ground, connect to pin 10 of SV203's port A (J3);
3N, 3E, 3S, 3W—signal wires, connect as shown Figure 6.15.

The location of the pins is shown in the datasheet of the compass (PDF). The output of the resistor ladder, Vout, can be connected either to pin 4 or pin 5 of SV203's port A (J3).

Determining Compass Heading

The encoded compass heading can be read by means of the AD4 or AD5 commands of the SV203 board, depending on whether Vout was connected to pin 4 or 5 of the analog input port A. The range of readings for each of the directions depends on the exact values of the resistors in the circuit, which vary due to manufacturing imprecision, and possibly to noise. The ranges we obtained were (these values may need adjustments for each particular set of resistors):

Heading	Low	High
North	149	151
Northeast	37	42
East	97	100
Southeast	78	82
South	197	202
Southwest	163	164
West	181	184
Northwest	115	117

6.7   ACCELEROMETERS

The suitability of accelerometers for mobile robot positioning was evaluated at the University of Michigan. In this informal study it was found that there is a very poor signal-to-noise ratio at lower accelerations (i.e., during low-speed turns). Accelerometers also suffer from extensive drift, and they are sensitive to uneven grounds, because any disturbance from a perfectly horizontal position will cause the sensor to detect the gravitational acceleration g. One low-cost inertial navigation system aimed at overcoming the latter problem included a tilt sensor. The tilt information provided by the tilt sensor was supplied to the accelerometer to cancel the gravity component projecting on each axis of the accelerometer. Nonetheless, the results obtained from the tilt-compensated system indicate a position drift rate of 1 to 8 cm/s (0.4 to 3.1 in/s), depending on the frequency of acceleration changes. This is an unacceptable error rate for most mobile robot applications.

6.8   GYROSCOPES

The mechanical gyroscope, a well-known and reliable rotation sensor based on the inertial properties of a rapidly spinning rotor, has been around since the early 1800s. The first known gyroscope was built in 1810 by G.C. Bohnenberger of Germany. In 1852, the French physicist Leon Foucault showed that a gyroscope could detect the rotation of the earth. In the following sections we discuss the principle of operation of various gyroscopes.

Anyone who has ever ridden a bicycle has experienced (perhaps unknowingly) an interesting characteristic of the mechanical gyroscope known as gyroscopic precession. If the rider leans the bike over to the left around its own horizontal axis, the front wheel responds by turning left around the vertical axis. The effect is much more noticeable if the wheel is removed from the bike, and held by both ends of its axle while rapidly spinning. If the person holding the wheel attempts to yaw it left or right about the vertical axis, a surprisingly violent reaction will be felt as the axle instead twists about the horizontal roll axis. This is due to the angular momentum associated with a spinning flywheel, which displaces the applied force by 90 degrees in the direction of spin. The rate of precession is proportional to the applied torque T:

$$T = I \, \Omega \tag{6.1}$$

where

T = applied input torque
I = rotational inertia of rotor
ω = rotor spin rate
Ω = rate of precession.

Gyroscopic precession is a key factor involved in the concept of operation for the north-seeking gyrocompass, as will be discussed later.

Friction in the support bearings, external influences, and small imbalances inherent in the construction of the rotor cause even the best mechanical gyros to drift with time. Typical systems employed in inertial navigation packages by the commercial airline industry may drift about 0.1^0 during a 6-hour flight.

6.8.1 Space-stable Gyroscopes

The earth's rotational velocity at any given point on the globe can be broken into two components: one that acts around an imaginary vertical axis normal to the surface, and another that acts around an imaginary horizontal axis tangent to the surface. These two components are known as the vertical earth rate and the horizontal earth rate, respectively. At the North Pole, for example, the component acting around the local vertical axis (vertical earth rate) would be precisely equal to the rotation rate of the earth, or 15^0/hr. The horizontal earth rate at the pole would be zero.

As the point of interest moves down a meridian toward the equator, the vertical earth rate at that particular location decreases proportionally to a value of zero at the equator. Meanwhile, the horizontal earth rate, (i.e., that component acting around a horizontal axis tangent to the earth's surface) increases from zero at the pole to a maximum value of 15^0/hr at the equator.

There are two basic classes of rotational sensing gyros: 1) rate gyros, which provide a voltage or frequency output signal proportional to the turning rate, and 2) rate-integrating gyros, which indicate the actual turn angle. Unlike the magnetic compass, however, rate-integrating gyros can only measure relative as opposed to absolute angular position, and must be initially referenced to a known orientation by some external means.

A typical gyroscope configuration is shown in Figure 6.16. The electrically driven rotor is suspended in a pair of precision low-friction bearings at either end of the rotor axle. The rotor bearings are in turn supported by a circular ring, known as the inner gimbal ring; this inner gimbal ring pivots on a second set of bearings that attach it to the outer gimbal ring. This pivoting action of the inner gimbal defines the horizontal axis of the gyro, which is perpendicular to the spin axis of the rotor as shown in Figure 6.16. The outer gimbal ring is attached to the instrument frame by a third set of bearings that define the vertical axis of the gyro. The vertical axis is perpendicular to both the horizontal axis and the spin axis.

Notice that if this configuration is oriented such that the spin axis points east-west, the horizontal axis is aligned with the north-south meridian. Since the gyro is space-stable (i.e., fixed in the inertial reference frame), the horizontal axis thus reads the horizontal earth rate component of the planet's rotation, while the vertical axis reads the vertical earth rate component. If the spin axis is rotated 90 degrees to a north-south alignment, the earth's rotation does not affect the gyro's horizontal axis, since that axis is now orthogonal to the horizontal earth rate component.

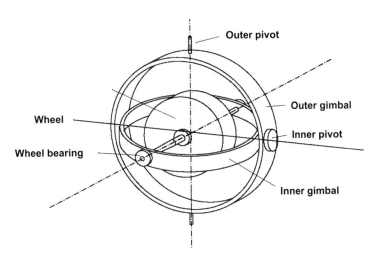

FIGURE 6.16 Typical two-axis mechanical gyroscope configuration (Everett, 1995).

6.8.2   Gyrocompasses

The gyrocompass is a special configuration of the rate-integrating gyroscope, employing a gravity reference to implement a north-seeking function that can be used as a true-north navigation reference. This phenomenon, first demonstrated in the early 1800s by Leon Foucault, was patented in Germany by Herman Anschutz-Kaempfe in 1903, and in the U.S. by Elmer Sperry in 1908. The U.S. and German navies had both introduced gyrocompasses into their fleets by 1911.

The north-seeking capability of the gyrocompass is directly tied to the horizontal earth rate component measured by the horizontal axis. As mentioned earlier, when the gyro spin axis is oriented in a north-south direction, it is insensitive to the earth's rotation, and no tilting occurs. From this it follows that if tilting is observed, the spin axis is no longer aligned with the meridian. The direction and magnitude of the measured tilt are directly related to the direction and magnitude of the misalignment between the spin axis and true north.

6.8.3   Gyros

Gyros have long been used in robots to augment the sometimes erroneous dead-reckoning information of mobile robots. Mechanical gyros are either inhibitively expensive for mobile robot applications, or they have too much drift. Work by Barshan and Durrant-Whyte aimed at developing an INS based on solid-state gyros, and a fiber-optic gyro was tested by Komoriya and Oyama.

Barshan and Durrant-Whyte

Barshan and Durrant-Whyte developed a sophisticated INS using two solid-state gyros, a solid-state triaxial accelerometer, and a two-axis tilt sensor. The cost of the complete system was £5,000 (roughly $8,000). Two different gyros were evaluated in this work. One was the ENV-O5S Gyrostar from [MURATA], and the other was the Solid State Angular Rate Transducer (START) gyroscope manufactured by [GEC]. Barshan and Durrant-Whyte evaluated the performance of these two gyros and found that they suffered relatively large drift, on the order of 5 to 15°/min. The Oxford researchers then developed a sophisticated error model for the gyros, which was subsequently used in an Extended Kalman Filter. Figure 6.17 shows the results of the experiment for the START gyro (left-hand side) and the Gyrostar (right-hand side). The thin plotted lines represent the raw output from the gyros, while the thick plotted lines show the output after conditioning the raw data in the EKF.

The two upper plots in Figure 6.17 show the measurement noise of the two gyros while they were stationary (i.e., the rotational rate input was zero, and the gyros should ideally show $\varphi = 0°/s$). Barshan and Durrant-Whyte determined that the standard deviation, here used as a measure for the amount of noise, was 0.16°/s for the START gyro and 0.24°/s for the Gyrostar. The drift in the rate output, 10 minutes after switching on, is rated at 1.35°/s for the Gyrostar (drift-rate data for the START was not given).

The more interesting result from the experiment in Figure 6.17 is the drift in the angular output, shown in the lower two plots. We recall that in most mobile robot applications one is interested in the heading of the robot, not the rate of change in the heading. The measured rate Æ must thus be integrated to obtain Æ. After integration, any small constant bias in the rate measurement turns into a constant-slope, unbounded error, as shown clearly in the lower two plots of Figure 6.17. At the end of the five-minute experiment, the START had accumulated a heading error of -70.8 degrees while that of the Gyrostar was -59 degrees (see thin lines in Figure 6.17). However, with the EKF, the accumulated errors

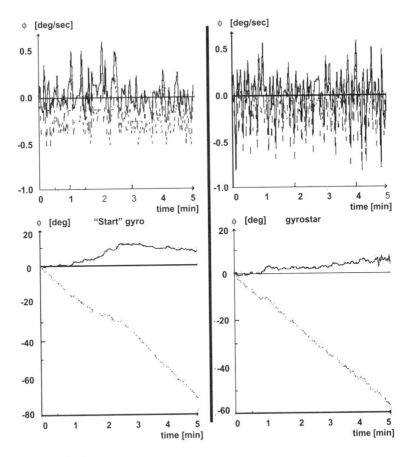

FIGURE 6.17

were much smaller: 12 degrees was the maximum heading error for the START gyro, while that of the Gyrostar was -3.8 degrees.

Overall, the results from applying the EKF show a five- to six-fold reduction in the angular measurement after a five-minute test period. However, even with the EKF, a drift rate of 1 to 3^0 /min can still be expected.

Komoriya and Oyama

Komoriya and Oyama conducted a study of a system that uses an optical fiber gyroscope, in conjunction with odometry information, to improve the overall accuracy of position estimation. This fusion of information from two different sensor systems is realized through a Kalman filter.

Figure 6.18 shows a computer simulation of a path-following study without (Figure 6.18a) and with (Figure 6.18b) the fusion of gyro information. The ellipses shows the reliability of position estimates (the probability that the robot stays within the ellipses at each estimated position is 90 percent in this simulation).

In order to test the effectiveness of their method, Komoriya and Oyama also conducted actual experiments with Melboy, the mobile robot shown in Figure 6.19. In one set of experiments, Melboy was instructed to follow the path shown in Figure 6.20a. Melboy's maximum speed was 0.14 m/s (0.5 ft./s) and that speed was further reduced at the corners of the path in Figure 6.20a. The final position errors without and with gyro information are compared and shown in Figure 6.20b for 20 runs. Figure 6.20b shows that the

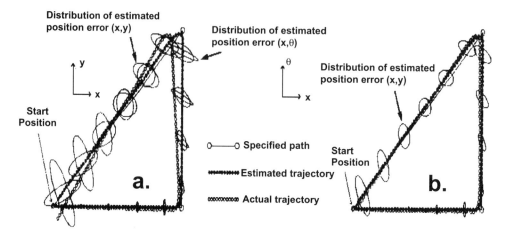

FIGURE 6.18   Computer simulation of a mobile robot run. a. Only odometry, without gyro information. b. Odometry and gyro information fused.

deviation of the position estimation errors from the mean value is smaller in the case where the gyro data was used (note that a large average deviation from the mean value indicates larger nonsystematic errors). Komoriya and Oyama explain that the noticeable deviation of the mean values from the origin in both cases could be reduced by careful calibration of the systematic errors of the mobile robot. We should note that from the description of this experiment it is not immediately evident how the "position estimation error" (i.e., the circles) in Figure 6.20b were found. In our opinion, these points should have been measured by marking the return position of the robot on the floor (or by any equivalent method that records the absolute position of

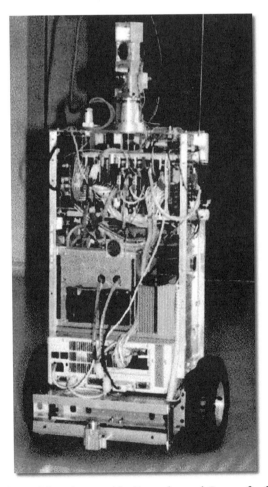

FIGURE 6.19   Melboy, the mobile robot used by Komoriya and Oyama for fusing odometry and gyro data.

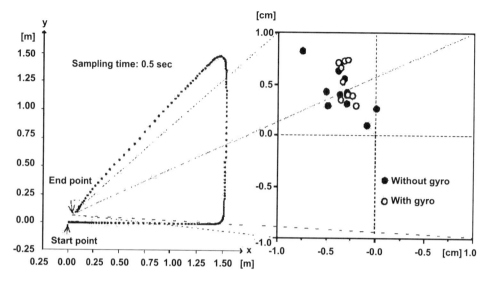

FIGURE 6.20 Experimental results from Melboy using odometry with and without a fiberoptic gyro. a. Actual trajectory of the robot for a triangular path. b. Position estimation errors of the robot after completing the path of a. Black circles show the errors.

the robot and compares it with the internally computed position estimation). The results of the plot in Figure 6.20b, however, appear to be too accurate for the absolute position error of the robot. In our experience an error on the order of several centimeters, not millimeters, should be expected after completing the path of Figure 6.20a. Therefore, we interpret the data in Figure 6.20b as showing a position error that was computed by the onboard computer, but not measured absolutely without gyro; white circles show the errors with the gyro.

6.9   LASER RANGE FINDER

A laser range finder is a device which uses a laser beam in order to determine the distance to a reflective object. The most common form of laser range finder operates on the time-of-flight principle by sending a laser pulse in a narrow beam toward the object and measuring the time taken by the pulse to be reflected off the target and returned to the sender. Due to the high speed of light, this technique is not appropriate for high-precision submillimeter measurements, where triangulation and other techniques are often used.

Operation

Pulse

The pulse may be coded in order to reduce the chance that the range finder can be jammed. It is possible to use Doppler effect techniques to judge whether the object is moving toward or away from the range finder, and if so, how fast.

The accuracy of the instrument is determined by the brevity of the laser pulse and the speed of the receiver. One that uses very short, sharp laser pulses and has a very fast detector can range an object to within a few centimeters.

Range

Despite the beam being narrow, it eventually spreads over long distances due to the divergence of the laser beam, as well as to scintillation and beam wander effects, caused by the presence of air bubbles in the air acting as lenses ranging in size from microscopic to roughly half the height of the laser beam's path above the earth.

These atmospheric distortions, coupled with the divergence of the laser itself and with transverse winds that serve to push the atmospheric heat bubbles laterally, may combine to make it difficult to get an accurate reading of the distance of an object, say, beneath some trees or behind bushes, or even over long distances of more than 1 km in open and unobscured desert terrain.

Some of the laser light might reflect off leaves or branches that are closer than the object, giving an early return and a reading which is too low. Alternatively, over distances longer than 1,200 ft. (365 m), the target, if in proximity to the earth, may simply vanish into a mirage, caused by temperature gradients in the air in proximity to the heated desert bending the laser light. All these effects have to be taken into account.

Discrimination

Some instruments are able to determine multiple returns, as above. These instruments use waveform-resolving detectors, which means they detect the amount of light returned over a certain time, usually very short. The waveform from a laser pulse that hit a tree and then the ground would have two peaks. The first peak would be the distance to the tree, and the second would be the distance to the ground.

The ability for aircraft-mounted instruments to see "through" dense canopies and other semireflective surfaces, such as the ocean, provide many applications for airborne instruments such as:

- Creating "bare earth" topographic maps—removing all trees
- Creating vegetation thickness maps

■   Measuring topography under the ocean
■   Forest fire hazard
■   Overwash threat in barrier islands

Applications

Military

In order to make laser range finders and laser-guided weapons less useful against military targets, various military arms may have developed laser-absorbing paint for their vehicles. Regardless, some objects don't reflect laser light very well and using a laser range finder on them is difficult.

3D Modelling

Laser range finders are used extensively in 3D object recognition, 3D object modelling, and a wide variety of computer vision-related fields. This technology constitutes the heart of the so-called time-of-flight 3D scanners. In contrast to the military instruments described above, laser range finders offer high-precision scanning abilities, with either single-face or 360-degree scanning modes.

A number of algorithms have been developed to merge the range data retrieved from multiple angles of a single object in order to produce complete 3D models with as little error as possible. One of the advantages that laser range finders offer over other methods of computer vision is that the computer does not need to correlate features from two images to determine depth information as in stereoscopic methods.

The laser range finders used in computer vision applications often have depth resolutions of tenths of millimeters or less. This can be achieved by using triangulation or refraction measurement techniques as opposed to the time-of-flight techniques used in LIDAR.

6.10   VISION-BASED SENSORS

Vision is our most powerful sense. It provides us with an enormous amount of information about the environment and enables rich, intelligent interaction in dynamic environments. It is therefore not surprising that a great deal of effort has been devoted to providing machines with sensors that mimic the capabilities of the human vision system. The first step in this process is the creation of the sensing devices that capture the same raw information light that the human vision system uses. The next section describes the two current technologies for creating vision sensors: CCD and CMOS. These sensors have specific limitations in performance when compared to the human eye, and it is important that the reader understand these limitations. Afterward, the second and third sections describe vision-based sensors that are commercially available, along with their disadvantages and most popular applications.

CCD and CMOS sensors

CCD Technology: The charged coupled device is the most popular basic ingredient of robotic vision systems today. The CCD chip is an array of light-sensitive picture elements, or pixels, usually with between 20,000 and several million pixels total. Each pixel can be thought of as a light-sensitive, discharging capacitor that is 5 to 25 micrometers in size. First, the capacitors of all pixels are charged fully, and then the integration period begins. As the photons of light strike each pixel, they liberate electrons, which are captured by electric fields and retained at the pixel. Over time; each pixel accumulates a varying level of charged based on the total number of photons that have struck it. After the period of integration is complete, the relative charges of all pixels need to be frozen and read. In a CCD the reading portion is performed at one corner of the CCD chip. The bottom row of pixel charges is transported to the corner and read, then the rows above shift down and the process is repeated. This means that each charge must be transported across the chip, and it is critical that the value be preserved. This requires specialized control circuitry and custom fabrication techniques to ensure the stability of transported charges.

The photodiodes used in the CCD chips (and CMOS chips as well) are not equally sensitive to all frequencies of light. They are sensitive to light in between 400 and 1000 nm wavelength. It is important to remember that photodiodes are less sensitive to the ultraviolet end of the spectrum (e.g., blue) and are overly sensitive to the infrared portion (e.g., heat).

You can see that the basic light-measuring process is colorless: it is just measuring the total number of photons that strike each pixel in the integration period. There are two common approaches for creating color images. If the pixels on the CCD chip are grouped into 2x2 sets of four, then red, green, and blue dyes can be applied to a color filter so that each individual pixel receives only light of one color. Normally, two pixels measure green while one pixel each measures red and blue light intensity. Of course, this one-chip color CCD has a geometric resolution disadvantage. The number of pixels in the systems has been cut effectively by a factor of four, and therefore the image resolution output by the CCD camera will be sacrificed.

The three-chip color camera avoids these problems by splitting the incoming light into three complete (lower intensity) copies. Three separate CCD chips receive the light, with one red, green, or blue filter over each entire chip. Thus, in parallel, each chip measures light intensity for one color, and the camera must combine the CCD chip's outputs to create a joint color image. Resolution is preserved in the solution, although the three-chip color cameras are, as one would expect, significantly more expensive and therefore rarely used in mobile robotics.

Both three-chip and single-chip color CCD cameras suffer from the fact that photodiodes are much more sensitive to the near-infrared end of the spectrum. This means that the overall system detects blue light much more poorly than red and green. To compensate, the gain must be increased on the blue channel, and this introduces the greater absolute noise on blue than on red and green. It is not uncommon to assume at least one to two bits of additional noise on the blue channel. Although there is no satisfactory solution to this problem today, over time the processes for blue detection has been improved and we expect this positive trend to continue.

The CCD camera has several camera parameters that affect its behavior. In some cameras, these values are fixed. In others, the values are constantly changing based on built-in feedback loops. In higher-end cameras, the user can modify the values of these parameters via software. The iris position and shutter speed regulate the amount of light being measured by the camera. The iris is simply a mechanical aperture that constricts incoming light, just as in standard 35 mm cameras. Shutter speed regulates the integration period of the chip. In higher end cameras, the effective shutter speed can be as brief at 1/30,000 seconds and as long as 2 seconds. Camera gain controls the overall amplification of the analog signal, prior to A/D conversion. However, it is very important to understand that, even though the image may appear brighter after setting high gain, the shutter speed and iris may not have changed at all. Thus gain merely amplifies the signal, and amplifies along with the signal all of the associated noise and error. Although useful in applications where imaging is done for human consumption (e.g., photography, television), gain is of little value to a mobile roboticist.

In color cameras, an additional control exists for white balance. Depending on the source of illumination in a scene (e.g., fluorescent lamps, incandescent lamps, sunlight, underwater filtered light, etc.), the relative measurements of red, green, and blue light that define pure white light will change dramatically. The human eye compensates for all such effects in ways that are not fully understood, but the camera can demonstrate glaring inconsistencies in which the same table looks blue in one image taken during the night, and yellow in another image taken during the day. White balance controls enable the user to change the relative gains for red, green, and blue in order to maintain more consistent color definitions in varying contexts.

The key disadvantages of CCD cameras are primarily in the areas of inconstancy and dynamic range. As mentioned above, a number of parameters can change the brightness and colors with which a camera creates its image. Manipulating these parameters in a way to provide consistency over time and over environments, for example, ensuring that a green shirt always looks green, and something dark gray is always dark gray, remains an open problem in the vision community.

The second class of disadvantages relates to the behavior of a CCD chip in environments with extreme illumination. In cases of very low illumination, each pixel will receive only a small number of photons. The longest possible integration period (i.e., shutter speed) and camera optics (i.e., pixel size, chip size, lens focal length, and diameter) will determine the minimum level of light for which the signal is stronger than random error noise. In cases of very high illumination, a pixel fills its well with free electrons and, as the well reaches its limit, the probability of trapping additional electrons falls and therefore the linearity between incoming light and electrons in the well degrades. This is termed saturation and can indicate the existence of a further problem related to cross-sensitivity. When a well has reached its limit, then additional light within the remainder of the integration period may cause further charge to leak into neighboring pixels, causing them to report incorrect values or even reach secondary saturation. This effect, called blooming, means that individual pixel values are not truly independent.

The camera parameters may be adjusted for an environment with a particular light level, but the problem remains that the dynamic range of a camera is limited by the well capacity of the individual pixel. For example, a high-quality CCD may have pixels that can hold 40,000 electrons. The noise level for reading the well may by 11 electrons, and therefore the dynamic range will be 40,000:11, or 3600:1, which is 35 dB.

CMOS Technology: The complementary metal oxide semiconductor chip is a significant departure from the CCD. It too has an array of pixels, but located alongside each pixel are several transistors specific to that pixel. Just as in CCD chips, all of the pixels accumulate charge during the integration period. During the data collection step, the CMOS takes a new approach: the pixel-specific circuitry next to every pixel measures and amplifies the pixel's signal, all in parallel for every pixel in the array. Using more traditional traces from general semiconductor chips, the resulting pixel values are all carried to their destinations.

CMOS has a number of advantages over CCD technology. First and foremost, there is no need for the specialized clock drivers and circuitry required in the CCD to transfer each pixel's charge down all of the array columns and across all of its rows. This also means that specialized semiconductor manufacturing processes are not required to create CMOS chips. Therefore the same production lines that create microchips can create inexpensive CMOS chips as well. The CMOS chip is so much simpler that it consumes significantly less power; incredibly, it operates with a power consumption that is one-hundredth the power consumption of a CCD chip. In a mobile robot, power is a scarce resource and therefore this is an important advantage.

On the other hand, the CMOS chip also faces several disadvantages. Most importantly, the circuitry next to each pixel consumes valuable real estate on the face of the light-detecting array. Many photons hit the transistors rather than the photodiode, making the CMOS chip significantly less sensitive than an equivalent CCD chip. Second, the CMOS technology is younger and, as a result, the best resolution that one can purchase in CMOS format continues to be far inferior to the best CCD chip available. Time will doubtless bring the high-end CMOS imagers closer to CCD imaging performance.

Given this summary of the mechanism behind CCD and CMOS chips, one can appreciate the sensitivity of any vision-based robot sensor to its environment. As compared to the human eye, these chips all have far poorer adaptation, cross-sensitivity, and dynamics range. As a result, vision sensors today continue to be fragile. Only over time, as the underlying performance of imaging chips improves, will significantly more robust vision-based sensors for mobile robots be available.

Camera output considerations: Although digital cameras have inherently digital output, throughout the 1980s and early 1990s, most affordable vision modules provided analog output signals, such as NTSC (National Television Standards Committee) and PAL (Phase Alternating Line). These camera systems included a D/A converter which, ironically, which would be counteracted on the computer using a framegrabber, effectively an A/D converter board situated, for example, on a computer's bus. The D/A and A/D steps are far from noisefree, and furthermore the color depth of the analog signal in such cameras was optimized for human vision, not computer vision.

More recently, both CCD and CMOS technology vision systems provide digital signals that can be directly utilized by the roboticist. At the most basic level, an imaging chip provides parallel digital I/O (input/output) pins that communicate discrete pixel level values. Some vision modules make use of these direct digital signals, which must be handled subject to hard-time constraints governed by the imaging chip. To relieve the real-time demands, researchers often place an image buffer chip between the imager's digital output and the computer's digital inputs. Such chips, commonly used in webcams, capture a complete image snapshot and enable non-real-time access to the pixels, usually in a single, ordered pass.

At the highest level, a roboticist may choose instead to utilize a higher-level digital transport protocol to communicate with an imager. Most common are the IEEE 1394 (firewire) standard and the USB (and USB 2.0) standards, although some order imaging modules also support serial (RS-232). To use any such high-level protocol, one most locate or create drive code both for that communication layer and for the particular implementation detail of the imaging chip. Take note, however, of the distinction between lossless digital video and the standard digital video stream designed for human visual consumption. Most digital video cameras provide digital output, but often only in compressed from. For vision researchers, such compression must be avoided as it not only discards information but even introduces image detail that does not actually exist, such as MPEG (Moving Picture Experts Group) discretization boundaries.

6.11   COLOR-TRACKING SENSORS

Although depth from stereo will doubtless prove to be a popular application of vision-based methods to mobile robotics, it mimics the functionality of existing sensors, including ultrasonic, laser, and optical range finders. An important aspect of vision-based sensing is that the vision chip can provide sensing modalities and cues that no other mobile robot sensor provides. One such novel sensing modality is detecting and tracking color in the environment.

Color represents an environmental characteristic that is orthogonal to range, and it represents both a natural cue and an artificial cue that can provide new information to a mobile robot. For example, the annual robot soccer events make extensive use of color both for environmental marking and for robot localization.

Color sensing has two important advantages. First, detection of color is a straightforward function of a single image; therefore no correspondence problem need be solved in such algorithms. Second, because color sensing provides a new, independent environmental cue, if it is combined (i.e., sensor fusion) with existing cues, such as data from stereo vision or laser range finding, we can expect significant information gains.

Efficient color-tracking sensors are now available commercially. Below, we briefly describe two commercial, hardware-based color-tracking sensors, as well as a publicly available software-based solution.

Cognachrome Color-tracking System

The Cognachrome Vision system from Newton Research Labs is a color-tracking hardware-based sensor capable of extremely fast color tracking on a dedicated processor. The system will detect color blobs based on three user-defined colors at a rate of 60 Hz. The cognachrome system can detect and report on a maximum of twenty-five objects per frame, providing centroid, bounding box, area, aspect ratio, and principal axis orientation information for each object independently.

This sensor uses a technique called constant thresholding to identify each color. In RGB (red, green, and blue) space, the user defines for each of R, G, and B a minimum and maximum value. The 3D box defined by these six constraints forms a color bounding box and any pixel with RGB values that are all within this bounding box is identified as a target. Target pixels are merged into large objects that are then reported to the user.

The cognachrome achieves a position resolution of one pixel for the centroid of each object in a field that is 200x250 pixels in size. The key advantage of this sensor, just as with laser range finding and ultrasonic, is that there is no load on the mobile robot's main sensor due to sensing modality, all processing is performed on sensor-specific hardware (i.e., a Motorola 68332 processor and a mated frame grabber). The cognechrome system costs several thousands dollars, but is being superseded by higher-performance hardware vision processors at Newton Labs, Inc.

CMUcam Robotic Vision Sensor: Recent advances in chip manufacturing, both in terms of CMOS imaging sensors and high-speed, readily available microprocessors at the 50+ MHz range, have made it possible to manufacture low-overhead intelligent vision sensors with functionality similar to cognachrome for a fraction of the cost. The CMUcam sensor is a recent system that mates a low-cost microprocessor with a consumer CMOS imaging chip to yield an intelligent, self-contained vision sensor for $100.

This sensor is designed to provide high-level information extracted from the camera image to an external processor that may, for example, control a mobile robot. An external processor configures the sensor's streaming data mode, for instance, specifying tracking mode for a bounded RGB or YUV value set. Then the vision sensor processes the data in real time and outputs a high level of information to the external consumer. At less than 150 mA of current draw, the sensor provides image color statistics and color-tracking services at approximately 20 frames per second at a resolution of 80X143.

CMVision Color-tracking Software Library

Because of the rapid speedup of processors in recent times, there has been a trend toward executing basic vision processing on a main processor within the mobile robot. Intel corporation's computer vision library is an optimized library for just such processing. In this sprit, the CMVision color-tracking software represents a state-of-the-art software solution for color tracking in dynamic environments. CMVision can track up to 32 colors at 30 Hz on a standard 200 MHz Pentium computer.

The basic algorithm this sensor uses is constant thresholding, as with Cognachrome, with the chief difference that the YUV color space is used instead of the RGB color space when defining a six-constraint bounding box for each color. While R, G, and B values encode the intensity of each color, YUV separate the color (or chrominance) measure from the brightness (or luminosity) measure. Y represents the image luminosity while U and V together captures its chrominance. Thus, a bounding box expressed in YUV space can achieve greater stability with respect to changes in illumination than is possible in RGB space.

The CMVision color sensors achieve a resolution of 160x120 and returns, for each object detected, a bounding box and a centroid. The software for CMVision is available freely with a Gnu public license.

Key performance bottlenecks for both the CMVision software, the CMU-cam hardware system, and the Cognachrome hardware system continue to be the quality of imaging chips and available computational speed. As significant advances are made on these frontiers one can expect packaged vision systems to witness tremendous performance improvements.

Making Infrared Sensors

Constructing the line sensor is straightforward. We would recommend that a prototype be made to test the operation of the sensor before committing to any permanent construction. Suggestions are presented below relating to sensor separation and placement, but you should experiment with your own configuration to see what works best.

Theory

The infrared emitter and detector sensors are shown below in Figure 6.21. You can buy these from a number of vendors, and depending on where they were purchased, the sensors shown in Figure 6.21 may differ in color and packaging from ones you're using. The following picture shows the most commonly found detector and emitter pair.

The circuit diagram is shown in Figure 6.22 and only one set of emitter/detector sensors is depicted. Pay close attention to the **anode (+ve)** and **cathode (GND)** of these sensors. Usually the longer leg is the anode (+ve) and the shorter leg is the cathode (GND) in most LEDs, including the infrared emitter. However, in the case of the detector it is the opposite, that is the longer LED is the cathode (GND) and the shorter leg is the anode (+ve).

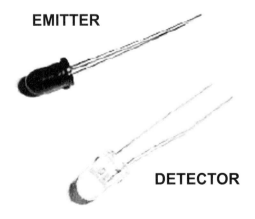

EMITTER

DETECTOR

FIGURE 6.21 **Infrared emitter and detector sensors.**

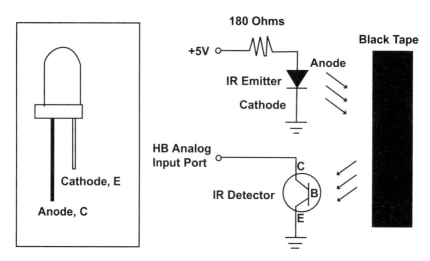

FIGURE 6.22   Circuit diagram for the infrared emitter/detector line sensor for the AT89c52 microcontroller.

The sensor for the line-following robot needs to be able to distinguish between the black tape and the white floor. IR sensors will be used instead of visible light sensors because visible light sensors are easily interfered with by ambient light and shadow. The longer wavelength of IR creates a stronger, more reliable signal while still being absorbed by the black tape.

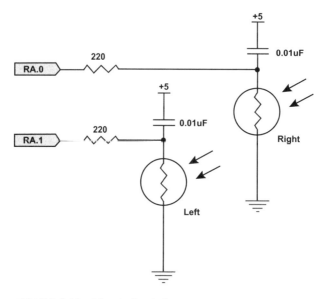

FIGURE 6.23   The IR circuit for 2 sensors.

A comparator will be used as a simple analog-to-digital converter to create a digital signal to send to the control stage of the robot. These sensors operate by either creating or removing a path to ground. When the IR transistor is not receiving a signal, the collector voltage is forced into the comparator. When the IR transistor receives a signal, a path to ground is created and a lower voltage is sent to the comparator.

Steps for Making the Infrared Sensor

The Interfacing of the Encoder

The next stage in the work is to obtain a feedback from the vehicle about its absolute position. This was calculated in the kinematics section by counting the speeds of the individual wheels and computing the position from that data. One disadvantage of the above method is that it assumes that the motor never misses a step. However, in a practical environment there may be a number of conditions when the robot misses a step. Hence, to measure the absolute and relative location of the WMR accurately, we need to obtain some feedback about the motion.

There are a number of ways for measuring the position of the WMR. One conventional way of determining their position in their environment is *odometry*. Odometry is the use of motion sensors to determine the robot's change in position relative to some known position. For example, if a robot is traveling in a straight line and if it knows the diameter of its wheels, then by counting the number of wheel revolutions it can determine how far it has traveled. Robots often have shaft encoders attached to their drive wheels, which emit a fixed number of pulses per revolution. By counting these pulses, the processor can estimate the distance traveled.

Odometry is the art of using this type of feedback to *estimate* where the robot is in its physical environment. It is strictly an estimate because of a number of physical limitations on the motors, gear boxes, interconnections, etc., that are involved, along with inertia, vary so greatly. Odometry is a very common position sensor for mobile robots, but it has its limitations. Since it is a cumulative measurement, any sensing error will increase as time passes. Robots may periodically need to use other sensors to precisely determine the robot's position to prevent excessive error buildup. Sources of odometry error are:

- Inaccurate wheel diameter measurement
- Different wheel sizes for multiple-wheel drive systems
- Pulse-counting errors in systems that use drive shaft encoders
- Slow odometry processing (considering only cumulative counts, not the dynamic count differences)

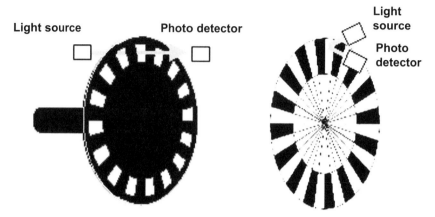

FIGURE 6.24   **Types of encoders.**

Shaft encoders are a way to measure angular changes in output shafts. A given shaft encoder, attached to the direct output wheel of a robot, generates a train of pulses to the **CPU** that indicates that the given wheel moved a number of degrees of rotation. When the wheel is turned in a given direction, the shaft encoders, in returning the information to the CPU, give the robot a sense of how much distance it is covering. Figure 6.24 presents two configurations in which incremental shaft encoders can read pulses. This will be discussed in detail in the next section.

Figure 6.25 shows how the photo detector sees the light and dark lines (or holes) and what the resulting voltage is.

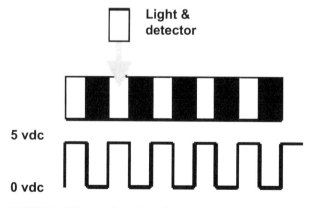

FIGURE 6.25   **Reading the pulses.**

The light and dark horizontal strips represent the bars (or holes) on the rotary encoder strip. Whenever a light strip is in front of the detector, the output voltage switches to high (5 V DC). Whenever a dark strip is in front, the voltage is low (0 V DC).

The processor reads these pulses and counts them (by various methods to be described later). By knowing how many pulses are detected per wheel revolution, and the circumference of the wheel, it can be calculated how far the robot has moved usually. But, there is one deficiency with this simple encoder. The processor will receive pulses that look the same whether the wheel is turning forward or reverse. Hence, the processor doesn't know whether to add or subtract the pulses from the distance traveled without some additional information. So this feature has to be added in the software, so that it may be able to tell which direction the wheel is turning (i.e., if the motor is being commanded forward, then the encoder pulses are contented forward), but using a more sophisticated encoder solves the problem directly.

The design and implementation of the shaft encoder will be done in the following steps:

1. Sensor mounting arrangement
2. Wheel discs design (for sensors to read RPM) and the circuitry
3. Software to read the pulses and implement in Turtle.

6.12   SENSOR MOUNTING ARRANGEMENT

The mounting of the sensor can be done in one of the following two types of arrangements (see Figure 6.26):

(1) Arrangement 1 – Light beam passing through slots

(2) Arrangement 2 – Light beam is reflected

In each case, a light beam is emitted and sensed by a photo detector. In arrangement 1, the light beam passes through slots in the sensing wheel; in arrangement 2, the light beam reflects from light and dark portions of the sensing wheel providing strong and weak reflections. Circuitry will take the signal sensed by the photo detector and turn it into a nice 0 to 5 V DC signal suitable for interfacing with a computer parallel port. This will be discussed in the next section.

Arrangement 1 is used in our system, Turtle. The shaft is attached to the disc. Each disc has alternating transparent and opaque segments. There is a light source on one side of the disc and a light detector on the other side. When

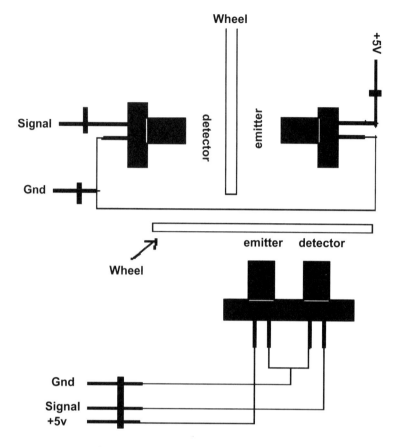

FIGURE 6.26   The arrangement of encoders.

a transparent segment of the disc lies between the source and detector the corresponding output is 1, and when an opaque sector lies between the source and detector the corresponding output is 0. Thus, the output alternates between 1 and 0 as the shaft is turned (see Figure 6.27).

6.13   DESIGN OF THE CIRCUITRY

Figure 6.28 represents the schematic representation of the circuit to be used to read pulses from the encoder. The pulse is read at the input pin no. 14. The pin attains the states 0 or 5 V when an opaque and transparent section crosses the receiver. By counting this change of states, the angular seed of the wheel can be computed. This is discussed in the next section.

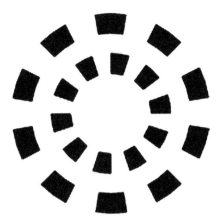

FIGURE 6.27   **Representation of the encoder wheel.**

6.14   READING THE PULSES IN A COMPUTER

After reading the pulse from the encoder, it is possible to count the pulses doing the polling. The software continuously samples an input pin (pin 14) with the detector signal on it and increments a counter when that signal changes state. However, it is difficult to do anything else with the software while you are doing this polling because a pulse may be missed while the software is off

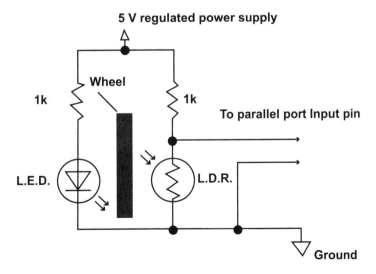

FIGURE 6.28   **The circuit for the encoder.**

doing something like navigation or controlling the motors. But, there is a better way. Many processors have interrupt capabilities. An interrupt is a hardware/software device that causes a software function to occur when something happens in the hardware. Specifically, whenever the detector A pulse goes high, the processor can be interrupted such that it suspends its ongoing navigation or motor control task, runs a special software routine (called an interrupt handler), which can compute the new distance traveled. When the interrupt handler is done, the processor automatically returns to the task it was working on when the interrupt occurred. The working program to count the encoder pulses is listed in Appendix II (b)

A simple description of the pulse counting process is presented here.
When the leading edge of a pulse occurs:
IF (motor command is forward) THEN distance = distance +1
IF (motor command is reverse) THEN distance = distance -1

If the motor command is not forward or reverse, distance is not changed. This avoids the possible problem of the robot stopping where the detector is right on the edge of an opaque section and might be tripping on and off with no real motion.

Another problem is that if the motor is rolling along and is commanded to zero, it might coast a little before stopping. One way to minimize this problem is to slowly decelerate to a stop so there is little or no coasting after the motor is set to zero.

Chapter 7   LEGGED ROBOTS

7.1 WHY STUDY LEGGED ROBOTS?

One need only watch a few slow-motion instant replays on the sports channels to be amazed by the variety and complexity of ways a human can carry, swing, toss, glide, and otherwise propel his body through space. Orientation, balance, and control are maintained at all times without apparent effort, while the ball is dunked, the bar is jumped, or the base is stolen, and such spectacular performance is not confined to the sports arena only. Behavior observable at any local playground is equally impressive from a mechanical engineering, sensory motor integration point of view. The final wonder comes when we observe the one-year-old infant's wobble with the knowledge that running and jumping will soon be learned and added to the repertoire.

Two-legged walking, running, jumping, and skipping are some of the most sophisticated movements that occur in nature, because the feet are quiet small and the balance at all times has to be dynamic; even standing still requires sophisticated control. If one falls asleep on ones feet he falls over. The human stabilizes the movement by integrating signals from:

- Vision, which includes ground position and estimates of the firmness of the ground and the coefficient of friction.
- Proprioception, that is, knowledge of the positions of all the interacting muscles, the forces on them and the rate of movement of the joints.
- The vesicular apparatus, the semicircular canals used for orientation and balance.

A very large number of muscles are used in a coordinated way to swing legs and the muscle in an engine consisting of a power source in series with an elastic connection. Various walking machines have been developed to imitate human legs, but none is as efficient as those of humans. Even the walking of four-legged animals is also highly complex and quite difficult to reproduce. The history of interest in walking machines is quite old. But until recently, they could not be developed extensively, because the high computational speed required by these systems was not available earlier. Moreover, the motors and power storage system required for these systems are highly expensive. Nevertheless, the high usefulness of these machines can discount on some of the cost factors and technical difficulty associated with the making of these systems. Walking machines allow locomotion in terrain inaccessible to other type of vehicles, since they do not need a continuous support surface, but the requirements for leg coordination and control impose difficulties beyond those encountered in wheeled robots. Some instances are in hauling loads over soft or irregular ground often with obstacles, agricultural operations, for movements in situations designed for human legs, such as climbing stairs or ladders. These aspects deserve great interest and, hence, various walking machines have been developed and several aspects of these machines are being studied theoretically.

In order to study them, different approaches may be adopted. One possibility is to design and build a walking robot and to develop study based on the prototype. An alternative perspective consists of the development of walking machine simulation models that serve as the basis for the research. This last approach has several advantages, namely lower development costs and a smaller time for implementing the modifications. Due to these reasons, several different simulation models were developed, and are used, for the study, design, optimization, and gait analysis and testing of control algorithms for artificial locomotion systems. The gait analysis and selection requires an appreciable modeling effort for the improvement of mobility with legs in unstructured environments. Several articles addressed the structure and selection of locomotion modes but there are different optimization criteria, such as energy efficiency, stability, velocity, and mobility, and its relative importance has not yet been clearly defined. We will address some of these aspects in these issues in the later sections of this chapter.

7.2   BALANCE OF LEGGED ROBOTS

The greatest challenge in building a legged robot is its balance. There are two ways to balance a robot body, namely static balance and dynamic balance. Both of these methods are discussed in this section.

7.2.1   Static Balance Methods

 Traditionally, stability in legged locomotion is taken to refer to static stability. The necessity for static stability in arthropods has been used as one of, if not the most important, reason why insects have at least six legs and use two sets of alternating tripods of support during locomotion. Numerous investigators have discussed the stepping patterns that insects require to maintain static stability during locomotion. Yet, few have attempted to quantify static stability as a function of gait or variation in body form. Research on legged walking machines provided an approach to quantify static stability. The minimum requirement to attain static stability is a tripod of support, as in a stool. If an animal's center of mass falls outside the triangle of support formed by its three feet on the ground, it is statically unstable and will fall. In the quasi-static gait of a robot or animal, the center of mass moves with respect to the legs, and the likelihood of falling increases the closer the center of mass comes to the edge of the triangle of support. In Figure 7.1 static balance is compared between six-legged and four-legged robotic platforms.

The problem of maintaining a stable platform is considerably more complex with four legs than it is with five, six, or more, since to maintain a statically stable platform there must always be at least three legs on the ground at any given time. Hence, with only four legs a shift in the center of mass is required to take a step. A six-legged robot, on the other hand, can always have a stable triangle—one that strictly contains the center of mass. In Figure 7.1 two successive postures or steps are shown for a four- and six-legged robot. In Figure 7.1 (a) the triangle for the first posture is stable because it contains the center of mass, but for the second posture the center of mass must be shifted in order for the triangle to be stable. In contrast, for the six-legged robot in Figure 7.1 (b) the center of mass can remain the same for successive postures.

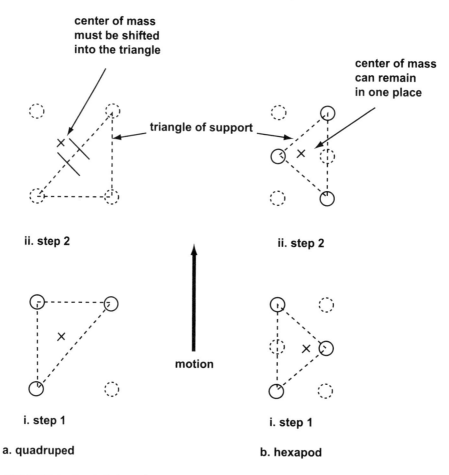

**center of mass
must be shifted
into the triangle**

**center of mass
can remain
in one place**

triangle of support

ii. step 2

ii. step 2

motion

i. step 1

i. step 1

a. quadruped

b. hexapod

FIGURE 7.1   Static balance in quadrupedal and hexapodal walking.

7.2.2   Dynamic Balance Methods

Dynamic stability analysis is required for all but the slowest movements. It was discovered that the degree of static stability decreased as insects ran faster, until at the highest speeds they became statically unstable during certain parts of each stride, even when a support tripod was present. Six- and eight-legged animals are best modeled as dynamic, spring-load, inverted pendulums in the same way as two- and four-legged runners. At the highest speeds, ghost crabs, cockroaches and ants exhibit aerial phases. In the horizontal plane, insects and other legged runners are best modeled by a dynamic, lateral leg spring, bouncing the animal from side to side. These models, and force and velocity measurements on animals, suggest that running at a constant average speed, while clearly a dynamical process, is essentially periodic in time. We define locomotor stability as the ability of characteristic measurements (i.e., state variables such as velocities, angles, and positions) to return to a steady state, periodic gait after a perturbation.

Quantifying dynamic stability—dynamical systems theory: The field of dynamical systems provides an established methodology to quantify stability. The aim of this text is not to explain the details of dynamical systems theory, but to give sufficient background so that those studying locomotion can see its potential in description and hypothesis formation. It is important to note that dynamical systems theory involves the formal analysis of how systems at any level of organization (neuron, networks, or behaviors) change over time. In this context, the term dynamical system is not restricted to a system generating forces (kinetics) and moving (kinematics), as is the common usage in biomechanics. The description of stability resulting from dynamical systems theory, which addresses mathematical models, differential equations, and iterated mappings, does not necessarily provide us with a direct correspondence to a particular biomechanical structure. Instead, the resulting stability analysis acts to guide our attention in productive directions to search for just such a link between coordination hypotheses from dynamical systems and mechanisms based in biomechanics and motor control.

Define and measure variables that specify the state of the system: The first task in the quantification of stability is to decide on what is best to measure. The goal is to specify a set of variables such as positions and velocities that completely define the state of the system. State variables are distinct from parameters such as mass, inertia, and leg length that are more or less fixed for a given animal. State variables change over time as determined by the dynamics of the system. Ideally, their values at any instant in time should allow the determination of all future values. Put another way, if two different trials of a running animal converge to the same values, their locomotion patterns should be very similar from that time forward.

Periodic trajectories called limit cycles characterize locomotion: During stable, steady-state locomotion, the value of state variables oscillates rhythmically over time (e.g., lateral velocity in Figure 7.2 A). In addition to representing the behavior of the state variables with respect to time, we can examine their behavior relative to one another. Figure 7.2 B shows a plot of the state variables (e.g., lateral, rotational, and fore-aft velocity) in state space. Time is no longer an axis, but changes as one moves along the loop in this three-dimensional space. The closed loop trajectory tells us that the system is periodic in time. Such a trajectory in state space is known as a limit cycle. If any other path converges to this cycle, it has stabilized to the same trajectory.

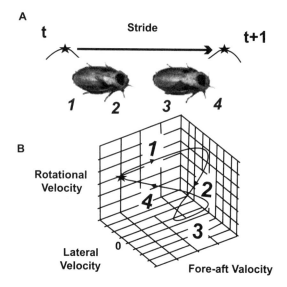

FIGURE 7.2   Periodic orbit or limit cycle. A. Variation in a single state variable, lateral velocity over one stride. A cycle is present within which lateral velocity repeats from t to t+1. B. Periodic orbit showing a limit cycle in state space. Lateral, rotation, and fore-aft velocity oscillate following a regular trajectory over a stride. Any point in the cycle can be considered an equilibrium point (star) of the associated return map.

Two types of stability exist—asymptotic and neutral. Characterizing stability requires perturbations to state variables (Figure 7.3). Most generally, stability can be defined as the ability of a system to return to a stable limit cycle or equilibrium point after a perturbation. There are at least two types of stable systems.

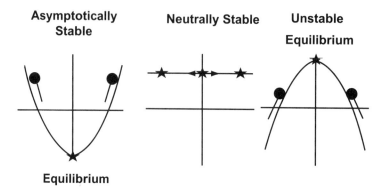

FIGURE 7.3   Types of stability; schematic representations of asymptotic stability with an equilibrium point (star), neutral stability with a continuum of equilibrium points, and an example of instability. The axes represent any two state variables.

In an asymptotically stable system, the return after the perturbation is to the original equilibrium or limit cycle. In a neutrally stable system, the return to stability after perturbation is to a new, nearby, equilibrium or limit cycle. In an unstable system perturbations tend to grow.

7.3   ANALYSIS OF GAITS IN LEGGED ANIMALS

Gait analysis is the process of quantification and interpretation of animal (including human) locomotion. Animal gaits have been studied throughout history, at least as far back as Aristotle. This section discusses some background material about the slower gaits, like creep, walking, and trotting, as well as some information about the faster gaits, such as running and galloping in four-legged animals. Trotting itself is not actually that slow, and some racehorses can trot almost as fast as others can gallop. However, trotting is similar enough to the walk that one might think a robot could be endowed with trotting ability as a natural extension of implementing the walk. We are not going to consider fast running as a viable means for robot mobility at this time.

The Creep

Creep, sometimes known as the crawl, is demonstrated by cats when stalking something—body low-slung to the ground, and slow meticulous movement of

FIGURE 7.4 **Cat displaying creep motion.**

only one leg at a time. We have also observed deer using this gait, when walking over broken ground. Compared to the cat, however, they keep their bodies fully erect, and lift each leg high during steps—to clear obstacles.

Tripod Stability: Whereas the alternating diagonal walk has dynamic stability, the creep has "static" stability. Only one leg is ever lifted from the ground at a time, while the other 3 maintain a stable tripod stance. The grounded legs are maintained in a geometry that keeps the center of mass of the body inside the triangle formed by the 3 points of the tripod at all times. As the suspended leg moves forward, the tripod legs shift the body forward in synchrony, so that a new stable tripod can be formed when the suspended leg comes down.

There are at least 2 variations of the creep:

1. The tripod can shift the body forward <u>simultaneously</u> with the suspended leg, giving a nice smooth forward movement. This method should provide good speed on level ground.
2. The tripod can shift the body forward <u>after</u> the suspended leg has touched down, giving a more tentative and secure forward movement. This method should be useful when engaging obstacles or moving over broken ground.

It seems there is little reason why a quadruped cannot be almost as stable as a hexapod, considering that a quad has 4 legs and it only takes 3 to build a stable tripod. Lift 1 leg for probing and stepping forward, and always keep 3 on the ground for stability. Just watch a clever cat negotiate the top of a fence.

Creep stability: The creep gait is "potentially" very stable, since 3 legs form a stable support tripod whenever any one leg is suspended.

However, lifting only 1 leg at a time sounds nice, but in the real world, this doesn't always work as predicted—for a quadruped, at least. It turns out, if the quad's legs are too short with respect to its body length, or they don't travel far enough (front-to-back) toward the midline of the body, or they are not coordinated well, then the 3 down legs may not form a stable tripod when the fourth is in the air. The down leg on the same side as the lifted leg, especially, must have its foot positioned far enough back, else the COG may not be contained within the stability triangle formed by the 3 down legs. Overall, creep stability relates to: body length, body width, leg length, leg angles, foot positions, and general distribution of weight on the body.

We have observed that deer do not have much problem with creep stability. Their legs are "very" long with respect to their body lengths, so keeping the COG within the stability tripod is easy.

Figure 7.5 illustrates how the static balance is maintained in creep gait. Given the position of the right front leg relative to the left rear, the associated edge of the stability triangle falls very close to the COG at this point. If those 2 legs are

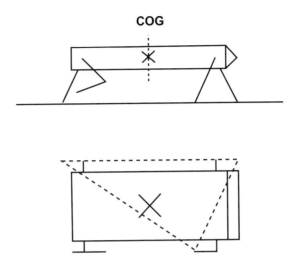

FIGURE 7.5 **COG during creep motion.**

not coordinated correctly, a point of instability may occur nearby in the stride. To improve stability here, the right front foot would have to touch down further back.

Walk

The dog in Figure 7.6 walks with a 4-time gait, LF (left-front), RR (right-rear), RF (right-front), LR (left-rear), then repeat. Presumably, most dogs prefer to start the walk with a front leg.

Notice that balance and support are maintained by the LR+RF "diagonal" while the LF and RR legs are suspended (positions 1, 2), and by the opposite diagonal for the other 2 legs (positions 5, 6). At the start of each step (positions 1, 5), the legs of the support diagonal are vertical, and the COG (center of gravity) of the dog is in the middle of the diagonal. Then the COG shifts forward as the stepping leg is extended (positions 2, 3, 6, 7), giving forward momentum to the body.

Regarding the suspended legs, the front leg precedes the rear leg (evident in positions 1–3 and 5–7) slightly, thus the 4-part cadence. Furthermore, during initiation of the succeeding steps (positions 4, 8), the front leg of the new step lifts slightly before the rear leg of the previous step touches down. This prevents the feet on the same side from banging into each other during the transition between diagonals, since for a normal stride; the rear pad comes down near the front pad mark.

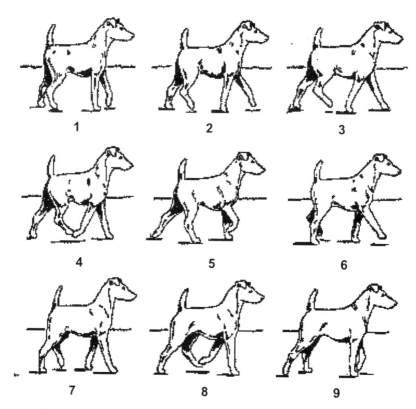

FIGURE 7.6 In the nine diagrams, the dog demonstrates the complete sequence of a full stride of the walk. The left front starts the action. Positions 1 and 2 show the right diagonal; 3, right diagonal and left front; 4, the left lateral; 5 and 6, the left diagonal; 7, the left diagonal and right front; 8, the right lateral; and 9 takes us back to the start.

The Trot

Faster gaits than the walk are the trot (positions 1–4) and the pace (positions 5, 6), which the German shepherd illustrates in Figure 7.7. The trot is basically a direct extension of the walk shown above, used for greater speed, while the pace is what a tired dog might use heading home after a long day on the range.

The trot is a two-time gait, LR+RF alternating with RR+LF. Like the walk above, the trot gets its stability by using alternating "diagonal bracing" under the torso (positions 1, 3), but here the legs work more closely in unison, the strides are longer and the forward lean is greater.

In practice, the phase of the front feet is slightly ahead of the rear, which keeps the feet on the same side from banging each other (positions 2, 4). Quite

FIGURE 7.7 Trot and pace gait.

obviously, the length of the stride with respect to the length of the torso and legs, as well as the precise phase relationship between front and rear legs on the same side, distinguish a successful trot from a bang, stagger, and stumble. Dogs with longer legs, compared to torso length, do not trot well.

The trot is an example of dynamic stability, where the animal could not hold its balance when stopped. However, because the trot always keeps the COG of the torso straddled by diagonal bracing, and the order of stepping is similar to that of the walk, one should think this gait would be easily implementable in a quadruped robot, once a successful walk has been achieved.

The Pace: In the pace, the dog uses "lateral" support, where both legs on each side work together, and the sides alternate. In the trot, the COG of the torso is not directly over the legs, but is centered within the diagonals. With the pacing gait, the COG will be offset from the supporting side unless the animal significantly leans its body sideways and angles its legs inward—an obvious stability problem.

Interestingly, humans also have something akin to a pacing gait. We have discovered empirically that, at the end of a long tiring hike, when struggling up the final hills, it turns out to be very comforting to twist and lean the body over onto the supporting leg with each step. This produces a kind of ducky waddling movement, which is not a good means for covering level ground at speed, but clearly takes some of the stress off tired muscles by shifting the COG directly over the grounded leg, so that the bones rather than the muscles take the effort. In a normal walk, the COG is normally kept in between the feet, so the leg muscles must take more of the effort. Because of the instability inherent with the pacing gait,

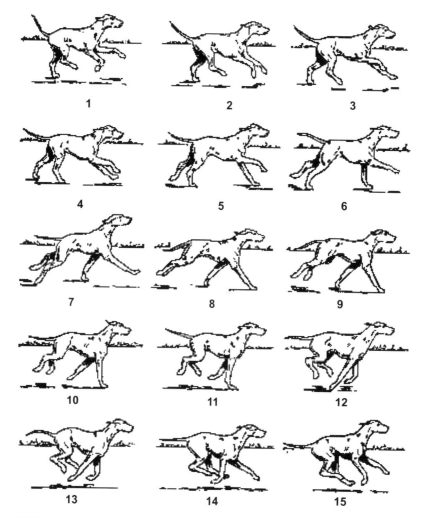

FIGURE 7.8   Galloping gait.

a pacing robot would probably fall over and go to sleep at the first opportunity. The trot, however, looks eminently doable in the robot.

Running Gaits

Galloping: The basic gallop is illustrated by the dog in Figure 7.8. This is a 4-step gait, shown here as LR, RR, LF, then RF. The leg ordering is very different from the walk and trot.

The RF leg is actually the "leading" leg here. The characteristic of this gait is that the leading leg bears the weight of the body over longer periods of time than any other leg, and is more prone to fatigue and injury. The single suspension phase (positions 13–15) is initiated by catapulting the entire body off the leading leg (positions 10–12). The force comes from the back legs pushing off onto the nonleading front leg, and then onto the leading leg (positions 3–8). Notice the length of time the back legs are suspended.

There is a position where all 4 legs are under the body, and others where either the 2 front or 2 rear legs are extended away from the body, but none where both fronts and both rears are extended simultaneously (as for the next gallop shown below). This gives a degree of rocking to the body, as the relative position of the COG moves forward and backward.

Flying: Lastly, the full-tilt double-suspension gallop is illustrated by the greyhound in Figure 7.9. (Suspension is the phase where all feet are off the ground simultaneously.)

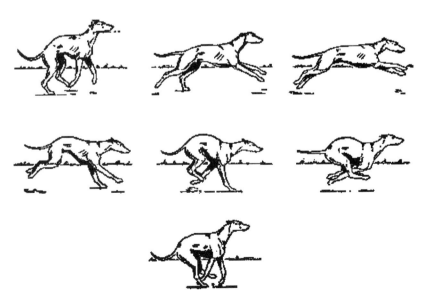

FIGURE 7.9   **Flying gait.**

The 3rd and 6th pictures illustrate well how the mirror image design of the legs produces symmetrical limb movement during running, and helps keep the COG of the animal at the same relative location during all aspects of the stride. From a stability viewpoint, this is clearly vital in preventing the dog's body from pitching nose over during the run.

However, implementing these running gaits in robots is very difficult. Stability is the biggest concern. Second is the availability of motors, which can generate such speed and power. The third problem is providing proper suspension to the entire body.

Robots can prosper from the aspects of animal dynamics, just described, in several ways:

- Robot bodies can be designed to take advantage of potential kinetic energy transformations, and especially forward inertia.
- Robot legs can be designed to absorb, store, and then rerelease the energy of foot impact.
- Robot legs can be arranged, like in the roach, to take advantage of self-stabilizing forces, which in turn can lessen the complexity of controllers and improve the overall stability of the devices in dynamic situations.

7.4   KINEMATICS OF LEG DESIGN

Kinematics of legged robots is similar to that of the robotic arm. The same kinematic analysis can be employed to legged robots to generate trajectories. So the kinematics can be divided into forward and inverse kinematics as discussed in Chapter 5 in Sections 5.3 and 5.4. We will implement forward and inverse kinematics to a problem of a general 3-DOF leg in the following sections.

7.4.1   Forward Kinematics

Let $\mu1, \mu2$, and $\mu3$ be angles for the rotator, shoulder joint, and knee joint, and let $l1$, $l2$, and $l3$ be the lengths of the shoulder, upper limb, and lower limb (Figure 7.10). Then the position (x, y, z) of the paw is represented by the following formulae:

$$x = l_3 \sin\theta_3 \cos\theta_1 + (l_2 + l_3 \cos\theta_3)\cos\theta_2 \sin\theta_1$$
$$y = l_1 \qquad\qquad + (l_2 + l_3 \cos\theta_3)\sin\theta_2,$$
$$z = l_3 \sin\theta_3 \sin\theta_1 - (l_2 + l_3 \cos\theta_3)\cos\theta_2 \cos\theta_1.$$

The frame assignment and calculation of joint link parameters is left for the user's practice. The users can revisit Section 5.3 (Forward Kinematics) for reference.

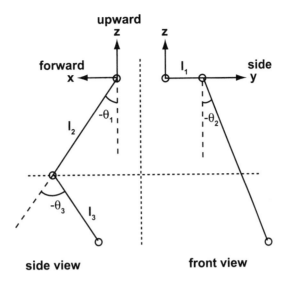

FIGURE 7.10   **Model and coordinate frame for leg kinematics presents the coordinate frame of a 3-DOF leg of a general 4-legged robot.**

7.4.2   Inverse Kinematics

We need the inverse kinematics in order to allow a high-level description of the gait, in which we design the movement of each leg by its position of paw, instead of these angles. Since our robot has 3 degrees of freedom, we have the following closed-form inverse kinematics:

$$\theta_3 = \pm \cos^{-1}\left(\frac{x^2 + z^2 + (y - l_1)^2 - (l_2^2 + l_2^3)}{2 l_2 l_3} \right),$$

$$\theta_2 = \sin^{-1}\left(\frac{y - l_1}{l_2 + l_3 \cos \theta_3} \right),$$

$$\theta_1 = \tan^{-1}\left(\frac{x}{-z} \right) m \cos^{-1}\left(\frac{(l_2 + l_3 \cos \theta_3) \cos \theta_2}{\sqrt{x^2 + z^2}} \right).$$

Note that we have two solutions for any reachable position (x, y, z), depending on if the angle $\mu 3$ of the knee is either positive or negative. Moreover, if $x = z = 0$ then $\mu 1$ can be arbitrarily chosen.

The derivation of inverse kinematics is left for the user's practice. The users can revisit Section 5.4 (Inverse Kinematics) for reference.

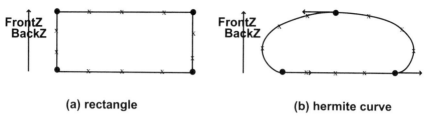

(a) rectangle   (b) hermite curve

FIGURE 7.11   Stroke shapes.

Leg Motion

The following parameters have to be specified to produce wheel-like leg motion.

Landing position **and** *leaving position*: positions at which the leg reaches the ground and leaves from it.

Lift height: lift heights of the leg.

Stroke shape: We can choose either rectangle or Hermite curve interpolation of three points, shown in Figure 7.11. A Hermite curve smoothly connects between specified points. These parameters determine the spatial trajectory of the paw. Moreover, an additional two parameters determine the position (x, y, z) of the paw at each time step.

Power ratio: time ratio of the paw touches the floor.

Time period: total steps of one stroke.

While the paw touches the floor, it moves on the straight line at a fixed speed.

7.5   DYNAMIC BALANCE AND INVERSE PENDULUM MODEL

There are several simplified models to describe leg dynamics. The simplest treats the body as an inverted pendulum mass, which transforms energy back and forth from gravitational potential energy at the top of the stance phase to kinetic energy during the lift phase of the step. In this model, the leg and body essentially rotate around the downed foot as a pivot point, with the up and down motions of the body mass related to the energy transformations. Inverse pendulum is a system consisting of a pole attached at its bottom to a moving cart. Whereas a normal pendulum is stable when hanging downward, a vertical inverted pendulum is inherently unstable, and must be actively balanced in order to remain upright, typically by moving the cart horizontally as part of a nonlinear feedback

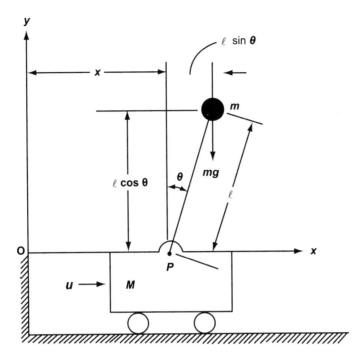

FIGURE 7.12   **Inverted pendulum on cart.**

system. We are going to discuss the dynamics of an inverse pendulum and its application to the balance of legged robots in this section. A sketch of this system is shown in Figure 7.12. The goal is to maintain the desired vertically oriented position at all times.

We will discuss here how the inverse pendulum model is applied to a biped to find out the dynamics of its motion. The biped locomotion is composed by two main phases: the single support phase and the double support phase. During the double support phase the robot is controllable, while it is not in the single support phase. This is due to the lack of an actuated foot. In the double support phase there is the transition of the support leg, from right to left or vice versa. Many authors have proposed a control strategy in which the double support phase is assumed to be instantaneous, and the actual biped locomotion is achieved by the single support phase. The main idea on which we rely for the control of the system is to reduce the dynamic of the system, in the single support phase, as that of a passive inverse pendulum. The single support phase starts when the robot is about to lift one of its two feet off the ground, more specifically the rear foot. This is the take-off configuration C_{TO}, which we have chosen to assign using three parameters: *dStep, dSwitch,* and

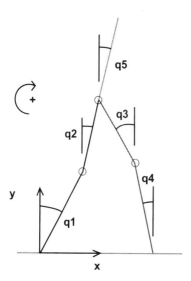

FIGURE 7.13   Degrees of freedom.

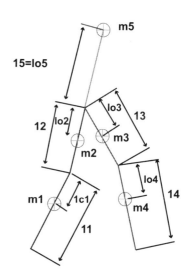

FIGURE 7.14   Dynamic parameters.

qMin. The meaning of these parameters is shown in Figure 7.13. In the C_{TO}, we treat the whole robot as an inverse pendulum (Figure 7.14). This is done by calculating the center of gravity *P* and the inertia momentum *I* of the whole system. Length of the pendulum *l* is given by the distance of the COG from the front foot. The initial position of the pendulum, given by the angle β is easily calculated.

The main principle of the control is to obtain a single support phase in which the inverse pendulum moves clockwise, in order to obtain the locomotion of the whole robot along the positive x-axis. To obtain such a goal, the initial velocity of the pendulum must be greater than a minimum value calculated using the conservation of the mechanical energy principle:

$$\dot{\beta}_0 = \sqrt{\frac{2mgl(1-\cos\beta_0)}{ml^2 + I}}$$

where β_0 is the initial velocity of the pendulum, m and I are the mass and the inertia momentum of the pendulum, g is the gravity acceleration, and l is the length of the pendulum.

Using the inverse kinematics, the velocity β_0 is translated in a set of given joint velocities in the C_{TO}; this configuration is now completely defined. We can now illustrate the basic principle of the control strategy: we want to use the double support phase to bring the robot from a given initial configuration C_0 to the

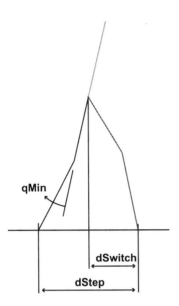

FIGURE 7.15   Take-off parameters.   Figure 7.16   Inverse pendulum scheme.

C_{TO}; in the C_{TO} the robot will switch to the single support phase where we will use the inverse pendulum approximation to control the dynamic of the swing leg. The objective is to swing this leg forward in coordination with the main body movement. Now we can explain more deeply the meaning of the various parameters used to define the C_{TO}. In Figures 7.15 and 7.16 the parameter *dStep* must be assigned accordingly to the robot legs length. A *dStep* too long would provoke unfavorable situations for the actuators, plus the body oscillation would be too great. A *dStep* too short would result in a slower locomotion and in stability problems. The parameter *qMin* is a security margin to avoid the singularity configuration of the knee joint, a situation in which the thigh and the leg are aligned. This parameter is directly related to the maximum leg extension, which is reached by the rear leg in the C_{TO}. The starting angle β_0 is directly related to the *dSwitch* parameter. A higher value for *dSwitch* causes a longer path for the inverse pendulum, thus a higher initial velocity, resulting in a higher actuators power output needed.

Dynamics of legged robots is an active field, where a lot of research is currently going on. Readers are advised to study detailed texts on legged robots to obtain a more detailed knowledge of the subject. A more practical interest in legged robots will start after the development of very low-weight and high-torque motors or other actuators and low-weight power storage devices.

Appendix A
Turtle.cpp

```
/*
(C)-2004,

Program for automatic tracking of Turtle.
*/

#include<iostream.h>
#include<conio.h>
#include<dos.h>
#include<math.h>
#include<fstream.h>
#include<iomanip.h>

// —Global Variables—
intF[4],Rl[4],Rr[4];
// —> change to enum boolean
int motor=1;
                              // F=1, Rl=2, Rr=3
int dir=0;                    // 0=clockwise, 1=anticlockwise
int pos_F,pos_Rl,pos_Rr;
                              // step position flags
ofstream outfile("Log.txt");
                              // Logfile
char stat_flag='L';
char motion;                  // L/l=logging, D/d=display, B/b=both, N/n=none
unsigned int F_count=0,Rl_count=0,Rr_count=0;
                              // step counts - initialized to 0
int F_pos[5000],Rl_pos[5000],Rr_pos[5000];
                              // step log array
// ——
```

```
// -Function Declarations-
void outSignal();
void initializeMotors();
void step(int motor, int dir);
void showMotorStatus();
void promptForStatus();
void showStepLog();
void write(int var);
void write(unsigned int var);
void write(float var);
void write(char* string);
void write(int var, int width);
void write(unsigned int var, int width);
void write(float var, int width);
// ——

void outSignal()
{
    int dec_378=0;
    for(int i=0;i<4;i++)
    {
            dec_378 += Rl[i]*pow(2,i);
    }
    for(i=4;i<8;i++)
    {
            dec_378 += Rr[i-4]*pow(2,i);
    }

    int dec_37A=0;
    for(i=0;i<4;i++)
    {
            dec_37A += F[i]*pow(2,i);
    }
    outportb(0x378,dec_378);
    outportb(0x37A,dec_37A);
    //write("\ndec_378=");write(dec_378);write
    (", dec_37A=");write(dec_37A);
}

void initializeMotors()
// To initialize the motors to its first position. add posflag
{
    F[0]=1;F[1]=0;F[2]=1;F[3]=1; pos_F=4;
    // Front
```

```
        Rl[0]=0;Rl[1]=1;Rl[2]=1;Rl[3]=0; pos_Rl=4;
        // Rear Left
        Rr[0]=0;Rr[1]=1;Rr[2]=1;Rr[3]=0; pos_Rr=4;
        // Rear Right
        outSignal();
}

void step(int motor, int dir)
{
    switch(motor)
    {
        case 1:
        {
            if(dir==0)
            {
                pos_F++;
                if(pos_F>4)
                        pos_F=1;
            }
            else
            {
                pos_F--;
                if(pos_F<1)
                        pos_F=4;
            }
            switch(pos_F)
            {
                case 1:
                {
                    F[0]=1;F[1]=0;F[2]=0;F[3]=0;break;
                }
                case 2:
                {
                    F[0]=0;F[1]=1;F[2]=0;F[3]=0;break;
                }
                case 3:
                {
                    F[0]=0;F[1]=1;F[2]=1;F[3]=1;break;
                }
                case 4:
                {
                    F[0]=1;F[1]=0;F[2]=1;F[3]=1;break;
                }
            }
```

```
                //F_pos[F_count++]=dir;
                break;
    }

    case 2:
    {
            if(dir==0)
            {
                    pos_Rl++;
                    if(pos_Rl>4)
                            pos_Rl=1;
            }
            else
            {
                    pos_Rl—;
                    if(pos_Rl<1)
                    pos_Rl=4;
            }
            switch(pos_Rl)
            {
                    case 1:
                    {
                            Rl[0]=0;Rl[1]=1;Rl[2]=0;Rl[3]=1;break;
                    }
                    case 2:
                    {
                            Rl[0]=1;Rl[1]=0;Rl[2]=0;Rl[3]=1;break;
                    }
                    case 3:
                    {
                            Rl[0]=1;Rl[1]=0;Rl[2]=1;Rl[3]=0;break;
                    }
                    case 4:
                    {
                            Rl[0]=0;Rl[1]=1;Rl[2]=1;Rl[3]=0;break;
                    }
            }
            //Rl_pos[Rl_count++]=dir;
            break;
    }

    case 3:
    {
            if(dir==1)
```

```
                    {
                            pos_Rr++;
                            if(pos_Rr>4)
                                    pos_Rr=1;
                    }
                    else
                    {
                            pos_Rr—;
                            if(pos_Rr<1)
                                    pos_Rr=4;
                    }
                    switch(pos_Rr)
                    {
                            case 1:
                            {
                                    Rr[0]=0;Rr[1]=1;Rr[2]=0;Rr[3]=1;break;
                            }
                            case 2:
                            {
                                    Rr[0]=1;Rr[1]=0;Rr[2]=0;Rr[3]=1;break;
                            }
                            case 3:
                            {
                                    Rr[0]=1;Rr[1]=0;Rr[2]=1;Rr[3]=0;break;
                            }
                            case 4:
                            {
                                    Rr[0]=0;Rr[1]=1;Rr[2]=1;Rr[3]=0;break;
                            }
                    }
                    //Rr_pos[Rr_count++]=dir;
                    break;
            }

    }
    outSignal();
}

void showMotorStatus()
            //overload << operator for single cout/outfile statement.
{
    write("\n————");
    write("\nF  - Pos:");write(pos_F);write(", Current seq:");
    for(int i=0;i<4;i++)
```

```
                write(F[i]);
    write("\nRl - Pos:");write(pos_Rl);write(", Current seq:");
    for(i=0;i<4;i++)
            write(Rl[i]);
    write("\nRr - Pos:");write(pos_Rr);write(", Current seq:");
    for(i=0;i<4;i++)
            write(Rr[i]);
    write("\n————");
}


void promptForStatus()
{
    /* do
    {
            cout<<"\nCircle (C) / Coordinate Points (P):";
            cin>>motion;
            flag=getche();
    }
    while(flag!='C'&&flag!='c'&&flag!='P'&&flag!='p'); */
    do
    {
            cout<<"\nLog staus (L) / Display onscreen (D) / Both (B) /
            None (N):";
            stat_flag=getche();
    }
    while(stat_flag!='L'&&stat_flag!='l'&&stat_flag!='D'&&stat_flag!
    ='d'&&stat_flag!='B'&&stat_flag!='b'&&stat_flag!='N'&&stat_flag!
    ='n');
}


void showStepLog()
{
    int i,j,k;
    write("\n–Step Log–");

    write("\nMotor F, steps=");write(F_count);write(":-\n");
    for(i=0;i<F_count;i++)
    {
            write(F_pos[i]);write(" ");
            if((i+1)%25==0)
                    write("\n");
    }

    write("\nMotor Rl, steps=");write(Rl_count);write(":-\n");
```

```cpp
    for(j=0;j<Rl_count;j++)
    {
            write(Rl_pos[j]);write(" ");
            if((j+1)%25==0)
                    write("\n");
    }

    write("\nMotor Rr, steps=");write(Rr_count);write(":-\n");
    for(k=0;k<Rr_count;k++)
    {
            write(Rr_pos[k]);write(" ");
            if((k+1)%25==0)
                    write("\n");
    }

    write("\n——");
}

void write(int var)
{
    if(stat_flag=='D'||stat_flag=='d'||stat_flag=='B'||stat_
    flag=='b')
            cout<<var;
    if(stat_flag=='L'||stat_flag=='l'||stat_flag=='B'||stat_
    flag=='b')
            outfile<<var;
}

void write(unsigned int var)
{
    if(stat_flag=='D'||stat_flag=='d'||stat_flag=='B'||stat_
    flag=='b')
            cout<<var;
    if(stat_flag=='L'||stat_flag=='l'||stat_flag=='B'||stat_
    flag=='b')
            outfile<<var;
}

void write(float var)
{
    if(stat_flag=='D'||stat_flag=='d'||stat_flag=='B'||stat_
    flag=='b')
            cout<<var;
    if(stat_flag=='L'||stat_flag=='l'||stat_flag=='B'||stat_
```

```
      flag=='b')
            outfile<<var;
}


void write(char* string)
{
   if(stat_flag=='D'||stat_flag=='d'||stat_flag=='B'||stat_
   flag=='b')
            cout<<string;
   if(stat_flag=='L'||stat_flag=='l'||stat_flag=='B'||stat_
   flag=='b')
            outfile<<string;
}


void write(int var, int width)    //width for setw manipulator
{
   if(stat_flag=='D'||stat_flag=='d'||stat_flag=='B'||stat_
   flag=='b')
            cout<<setw(width)<<var;
   if(stat_flag=='L'||stat_flag=='l'||stat_flag=='B'||stat_
   flag=='b')
            outfile<<setw(width)<<var;
}


void write(unsigned int var, int width)
{
   if(stat_flag=='D'||stat_flag=='d'||stat_flag=='B'||stat_
   flag=='b')
            cout<<setw(width)<<var;
   if(stat_flag=='L'||stat_flag=='l'||stat_flag=='B'||stat_
   flag=='b')
            outfile<<setw(width)<<var;
}


void write(float var, int width)
{
   if(stat_flag=='D'||stat_flag=='d'||stat_flag=='B'||stat_
   flag=='b')
            cout<<setw(width)<<var;
   if(stat_flag=='L'||stat_flag=='l'||stat_flag=='B'||stat_
   flag=='b')
            outfile<<setw(width)<<var;
}
```

```cpp
#include<graphics.h>
#include<stdio.h>

// —Global Variables—
const int screen_w=640,screen_h=480;
//the resolution of the graphics mode. values 640x480 for VGAHI driver
float q1_SF=4,q2_SF=4;
//the default scaling factors in the q1 & q2 co-ordinate axis
respectively. can change in the main prog
int q1_offset=320,q2_offset=240;
                        //should be a portin of screen_w & screen_h
int box1_x=10,box1_y=395,box1_w=150,box1_h=75;
//defines the position of the display box 1
int box2_x=330,box2_y=450,box2_w=300,box2_h=20;
//defines the position of the display box 2
int position_update_int=4;
//specifies intervals after which intermediate co-ords are displayed on
screen
// ——

// —Function Declarations—
void initializeGFX();
void initializeGFX(float _q1_SF, float _q2_SF);
//initialize the plot
void closeGFX();
void plot(float _q1, float _q2, int flag);
//flag specifies the type of entity to be plotted. 0=point, 1=small
filled circle.
void displayPar(float _delta, float _v_a);
void displayPosition(float _q1, float _q2, float _q3);
void displayStatus(char* string);
int q1ToScreenx(float _q1);
int q2ToScreeny(float _q2);
float screenxToq1(int x);
float screenyToq2(int y);
// ——

void initializeGFX()
{
    //registerbgidriver(EGAVGA_driver);
    int driver=DETECT,mode=VGAHI;
    //q1_SF=_q1_SF;
    //q2_SF=_q2_SF;
    initgraph(&driver, &mode, "");
```

```
//3rd parameter specifies dir of gfx drivers. update path.
setcolor(EGA_WHITE);
setlinestyle(SOLID_LINE,0,NORM_WIDTH);
line(screen_w/2,0,screen_w/2,screen_h-1);
line(0,screen_h/2,screen_w-1,screen_h/2);
setfillstyle(SOLID_FILL,EGA_BLACK);
bar3d(box1_x,box1_y,box1_x+box1_w,box1_y+box1_h,0,0);
line(box1_x,box1_y+55,box1_x+box1_w,box1_y+55);
bar3d(box2_x,box2_y,box2_x+box2_w,box2_y+box2_h,0,0);
const int marker_w=screen_w/14;
const int marker_h=screen_h/14;
int x,y;
char ch[10];
settextjustify(CENTER_TEXT,TOP_TEXT);
x=screen_w/2;
y=screen_h/2;
while(x<screen_w)
{
        x+=marker_w;
        line(x,y-2,x,y+2);
        sprintf(ch,"%i",int(screenxToq1(x)));
        outtextxy(x,y+5,ch);
}
x=screen_w/2;
while(x>=0)
{
        x-=marker_w;
        line(x,y-2,x,y+2);
        sprintf(ch,"%i",int(screenxToq1(x)));
        outtextxy(x,y+5,ch);
}
settextjustify(RIGHT_TEXT,CENTER_TEXT);
x=screen_w/2;
while(y<screen_h)
{
        y+=marker_h;
        line(x-2,y,x+2,y);
        sprintf(ch,"%i",int(screenyToq2(y)));
        outtextxy(x-5,y,ch);
}
y=screen_h/2;
while(y>=0)
{
        y-=marker_h;
```

```
        line(x-2,y,x+2,y);
        sprintf(ch,"%i",int(screenyToq2(y)));
        outtextxy(x-5,y,ch);
    }
}

void initializeGFX(float _q1_SF, float _q2_SF)
{
    q1_SF=_q1_SF;
    q2_SF=_q2_SF;
    initializeGFX();
}

void closeGFX()
{
    getch();
            //closes after keypress
    closegraph();
}

void plot(float _q1, float _q2, int flag)
//copies _q1 & _q2 of q1 & q2 used.
{
    int q1_p=_q1*q1_SF+q1_offset;
    //valid values of (q1_p,q2_p) range from 0,0 to 639,479 in VGAHI.
    int q2_p=screen_h-(_q2*q2_SF+q2_offset);
    switch(flag)
    {
            case 0:
            {
                    putpixel(q1_p,q2_p,EGA_LIGHTGREEN);
                    break;
            }

            case 1:
            {
                    setfillstyle(SOLID_FILL,EGA_LIGHTRED);
                    fillellipse(q1_p,q2_p,3,3);
                    break;
            }
    }
}

void displayPar(float _delta_a, float _v_a)
```

```
{
    char ch[50];
    settextjustify(LEFT_TEXT,TOP_TEXT);
    setfillstyle(SOLID_FILL,EGA_BLACK);
    bar(box1_x+1,box1_y+1,box1_x+box1_w-1,box1_y+54);
    sprintf(ch,"delta_a = %5.3f",_delta_a);
    outtextxy(box1_x+5,box1_y+5,ch);
    sprintf(ch,"v_a = %6.2f",_v_a);
    outtextxy(box1_x+5,box1_y+20,ch);
}


void displayPosition(float _q1, float _q2, float _q3)
{
    char ch[50];
    settextjustify(LEFT_TEXT,TOP_TEXT);
    setfillstyle(SOLID_FILL,EGA_BLACK);
    bar(box1_x+1,box1_y+56,box1_x+box1_w-1,box1_y+box1_h-1);
    sprintf(ch,"%5.1f,%5.1f,%5.1f",_q1,_q2,_q3);
    outtextxy(box1_x+5,box1_y+60,ch);
}


void displayStatus(char* string)
{
    settextjustify(LEFT_TEXT,TOP_TEXT);
    setfillstyle(SOLID_FILL,EGA_BLACK);
    bar(box2_x+1,box2_y+1,box2_x+box2_w-1,box2_y+box2_h-1);
    outtextxy(box2_x+5,box2_y+5,string);
}


int q1ToScreenx(float _q1)
{
    return int(_q1*q1_SF+q1_offset);
}


int q2ToScreeny(float _q2)
{
    return int(screen_h-(_q2*q2_SF+q2_offset));
}


float screenxToq1(int x)
{
    return (x-q1_offset)/q1_SF;
}
```

```cpp
float screenyToq2(int y)
{
    return (screen_h-q2_offset-y)/q2_SF;
}


#include<process.h>

// —Global Variables—
const float step_angle=(0.015865),step_distance=0.74;
// units in radians and mm —>may be diff for motors
const float l=113.5,b=137.5,r=47;
// l=length, 2b=width, in mm
float v=0,delta=0;
// current centroidal speed(mm/s) and steering angle(radians)
float q1=0,q2=0,q3=0;
// instantaneous global co-ordinates of vehicle CoG q1,q2 in mm and q3
in radians
float v_Rl,v_Rr;
// theoretical velovity of rear left and right wheel
float v_Rl_a=0,v_Rr_a=0;
// actual achievable velocities
float v_a=0,delta_a=0,delta_p=0;
// actual achievable centroidal velocity and steering angle
const int steering_delay=10;
// delay for steering steps, front motor, in ms
int interval_count=0;            // interval counter
const float delta_a_min=0.01;
// min value of delta_a for making a finite radius
float q1_int[800],q2_int[800],q3_int[800];
// instantaneous global co-ordinates in the time interval - for plotting
purposes only
// ——


// —Function Declarations—
int round(float num);
void moveDist(float dist, float speed, int motor, int dir);
void moveAngle(float angle, float omega, int motor, int dir);
void setSteering(float _delta);
void moveVehicle(float _delta, float _v, float _t, float _distance);
void moveVehicle1(float _delta, float _v, float _t, float _distance);


// ——


int round(float num)
```

```
{
    if((num-floor(num))<0.5)
            return floor(num);
    else
            return ceil(num);
}


void moveDist(float dist, float speed, int motor, int dir)
{
    int steps=int(dist/step_distance);
    // dist is integral multiple of step_distance
            int time_delay=int(1000/(speed/step_distance));
    // time delay is in ms
    write("\nMotor ID=");write(motor);write(",
    steps=");write(steps);write(", time_delay=");write(time_delay);
    write(", dir=");write(dir);
    for(int i=0;i<steps;i++)
    {
            step(motor,dir);
            delay(time_delay);
    }
}


void moveAngle(float angle, float omega, int motor, int dir)
{
    int steps=int(angle/step_angle);
    // angle is integral multiple of step_angle
            int time_delay=int(1000/(omega/step_angle));
    // time delay in ms
    write("\nMotor ID=");write(motor);write(", steps=");
    write(steps);write(", time_delay=");write(time_delay);
    write(", dir=");write(dir);
    for(int i=0;i<steps;i++)
    {
            step(motor,dir);
            delay(time_delay);
    }
}


void moveVehicle1(float _delta, float _v, float _t, float _distance)
{
    int dir;
    float x,y;
    // x & y increments in LCS per interval
```

```
interval_count++;
int p=0;
write("\n-Interval ");write(interval_count);write(" Step Log-");

if(_v>=0)
        dir=0;
else
        dir=1;

if(_t==-1)
        _t=_distance/_v;

//Calculating v_a, v_Rl_a & v_Rr_a
delta_a=_delta;
v_Rl=_v*(1-(b/l)*tan(delta_a));
v_Rl_a=(round((v_Rl*_t)/step_distance)*step_distance)/_t;
v_Rr_a=v_Rl_a*(1+(b/l)*tan(delta_a))/(1-(b/l)*tan(delta_a));
v_a=v_Rl_a/(1-(b/l)*tan(delta_a));
write("\nv_a=");write(v_a);write(",
v_Rl_a=");write(v_Rl_a);write(", v_Rr_a=");write(v_Rr_a);
//----

displayPar(delta_a,v_a);
displayPosition(q1,q2,q3);

//Calculating intermediate positions
float t_int=_t/100;
float x_int,y_int;
q1_int[0]=q1;q2_int[0]=q2;q3_int[0]=q3;
for(p=0;p<100;p++)
{
        //if global array used for whole problem then we've to use
        a global increment (ie p here).
        if(fabs(delta_a)>delta_a_min)
        {
                x_int=(l/tan(delta_a))*sin((v_a/l)*tan(delta_a)*t_int);
                y_int=(l/tan(delta_a))*(1-cos((v_a/
                l)*tan(delta_a)*t_int));
                q1_int[p+1]=q1_int[p]+(x_int*cos(q3_int[p])-
                y_int*sin(q3_int[p]));
                q2_int[p+1]=q2_int[p]+(x_int*sin(q3_int[p])+
                y_int*cos(q3_int[p]));
                q3_int[p+1]=q3_int[p]+((v_a/l)*tan(delta_a)*t_int);
                //not required
```

```
                    }
                    else
                    {
                            x_int=v_a*t_int;
                            y_int=0;
                            q1_int[p+1]=q1_int[p]+(x_int*cos(q3_int[p])-
                            y_int*sin(q3_int[p]));
                            q2_int[p+1]=q2_int[p]+(x_int*sin(q3_int[p])+
                            y_int*cos(q3_int[p]));
                            q3_int[p+1]=q3_int[p]+0;
                            //not required
                    }


            }

    if(fabs(delta_a)>delta_a_min)
    {
            x=(l/tan(delta_a))*sin((v_a/l)*tan(delta_a)*_t);
            y=(l/tan(delta_a))*(1-cos((v_a/l)*tan(delta_a)*_t));
            q1+=x*cos(q3)-y*sin(q3);
            q2+=x*sin(q3)+y*cos(q3);
            q3+=((v_a/l)*tan(delta_a)*_t);
    }
    else
    {
            x=v_a*_t;
            y=0;
            q1+=x*cos(q3)-y*sin(q3);
            q2+=x*sin(q3)+y*cos(q3);
            q3+=0;
    }
    write("\nGlobal co-ordinates of vehicle CoG:
    [");write(q1,4);write(q2,4);write(",");write(q3,4);write("]");
    //——

    //Calculating step timing array
    if((((fabs(v_Rl_a)*_t)/step_distance)<1.002)||(((fabs(v_Rr_a)*_t)/
    step_distance)<1.002))
    {
            displayStatus("Parameters out of range. Exiting...");
            write("\nParameters out of range. Exiting...");
            closeGFX();
            exit(1);
            /*if(v_Rl_a<=v_Rr_a)
```

```
            v_Rl_a=(round(1.02)*step_distance)/_t;*/
}


unsigned int delay_Rl=(_t/(((fabs(v_Rl_a)*_t)/step_distance)-
1))*1000;
//delay in ms ->exit on delay=0 ->check for single step
unsigned int delay_Rr=(_t/(((fabs(v_Rr_a)*_t)/step_distance)-
1))*1000;
//int data type limits max value of delay to 65,534ms

unsigned int timer_Rl[1000],timer_Rr[1000],timer_main[2000][2];
//step timing array for Rl, Rr, and union of the two.
timer_Rl[0]=0;
//the second index in timer_main denotes the motor id to be stepped.
timer_Rr[0]=0;
int i=0,j=0;

while((delay_Rl*(i+1))<=(_t*1000))
{
        i++;
        timer_Rl[i]=delay_Rl*i;
}
while((delay_Rr*(j+1))<=(_t*1000))
{
        j++;
        timer_Rr[j]=delay_Rr*j;
}

int k=1,l=1,m=0,flag;
timer_main[m][0]=0; timer_main[m][1]=5;
for(k=1;k<=i;k++)
{
        flag=1;
        while((l<=j)&&(flag==1))
        {
                if(timer_Rl[k]<timer_Rr[l])
                {
                        timer_main[++m][0]=timer_Rl[k];
                        timer_main[m][1]=2;
                        flag=0;
                }
                else
                {
                        if(timer_Rl[k]==timer_Rr[l])
```

```
                                        {
                                                timer_main[++m][0]=timer_Rl[k];
                                                timer_main[m][1]=5;
                                                l++;
                                                flag=0;
                                        }
                                        else
                                        {
                                                timer_main[++m][0]=timer_Rr[l];
                                                timer_main[m][1]=3;
                                                l++;
                                        }
                                }
                        }

                /*if(l>j)                                   //still not tested
                {
                        timer_main[++m][0]=timer_Rl[k];
                        timer_main[m][1]=2;
                }*/
        }
        for(l;l<=j;l++)
        {
                if(timer_Rl[i]!=timer_Rr[l])
                {
                        timer_main[++m][0]=timer_Rr[l];
                        timer_main[m][1]=3;
                }
        }

        if(stat_flag=='D'||stat_flag=='d'||stat_flag=='B'||stat_
        flag=='b')
        {
                cout<<"\nmotor Rl - "<<(i+1)<<" steps";
                cout<<"\nmotor Rr - "<<(j+1)<<" steps";
        }
        if(stat_flag=='L'||stat_flag=='l'||stat_flag=='B'||stat_
        flag=='b')
        {
                outfile<<"\nmotor Rl - "<<(i+1)<<" steps";/*cout<<" with
                step times:-\n";
                for(p=0;p<=i;p++)
                {
                        outfile<<setw(4)<<timer_Rl[p]<<"|";
```

```cpp
                        if((p+1)%25==0)
                                outfile<<"\n";
                }*/
                outfile<<"\nmotor Rr - "<<(j+1)<<" steps";/*cout<<" with
                step times:-\n";
                for(p=0;p<=j;p++)
                {
                        outfile<<setw(4)<<timer_Rr[p]<<"|";
                        if((p+1)%25==0)
                                outfile<<"\n";
                }*/
                /*outfile<<"\ncombined step timing array:-\n";
                for(p=0;p<=m;p++)
                {
                        outfile<<setw(4)<<timer_main[p][0]<<"("<<timer_main
                        [p][1]<<")|";
                        if((p+1)%25==0)
                                outfile<<"\n";
                }*/
        }
write("\n————————");
//——

//Executing step sequence
displayStatus("Moving Vehicle");
step(2,dir);
step(3,dir);
p=0;
for(int n=1;n<=m;n++)
{
        delay(timer_main[n][0]-timer_main[n-1][0]);

        if(timer_main[n][1]==5)
        {
                step(2,dir);
                step(3,dir);
        }
        else
        {
                step(timer_main[n][1],dir);
        }

        while(((t_int*1000*p)>=timer_main[n-1][0])
        &&((t_int*1000*p)<=timer_main[n][0]))
```

```
                {
                        plot(q1_int[p],q2_int[p],0);
                        displayPosition(q1_int[p],q2_int[p],q3_int[p]);
                        p++;
                }

        }
        //——
        plot(q1,q2,0);
}

void setSteering(float _delta)
{
    float delta_a_inc, delta_a_inc_cur=0;
    delta_a_inc=(round(_delta*(b/r)/step_angle))*step_angle;
    //delta_a =delta;
    delta_p=(round(_delta/step_angle))*step_angle;
    //round off in terms of step_angle

    write("\ndelta_a=");write(delta_a);
    write(", delta_a_increment=");write(delta_a_inc);
    write(", motor F-steps=");write(abs(round(delta_a_inc/
    step_angle)));write(", with
    delay=");write(steering_delay);write("ms");
    if(delta_a_inc>0)
            write(", in dir=0");
    if(delta_a_inc<0)
            write(", in dir=1");
    displayStatus("Steering");

    if(delta_a_inc>0)
    {
            while(delta_a_inc_cur<delta_a_inc)
            {
                    step(2,0);step(3,1);
                    delay(steering_delay);
                    delta_a_inc_cur+=step_angle;
            }
    }
    if(delta_a_inc<0)
    {
            while(delta_a_inc_cur>delta_a_inc)
            {
                    step(2,1);step(3,0);
```

```cpp
                        delay(steering_delay);
                        delta_a_inc_cur-=step_angle;

                }
        }
}

void moveVehicle(float _delta, float _v, float _t, float _distance)
{
    int dir;
    float x,y;
    // x & y increments in LCS per interval
    interval_count++;
    int p=0;
    write("\n—Interval ");write(interval_count);write(" Step Log–");

    if(_v>=0)
            dir=0;
    else
            dir=1;

    if(_t==-1)
            _t=_distance/_v;

    setSteering(_delta);
    q3+=delta_p;
    delta_a=0;

    //Calculating v_a, v_Rl_a & v_Rr_a
    v_Rl=_v*(1-(b/l)*tan(delta_a));
    v_Rl_a=(round((v_Rl*_t)/step_distance)*step_distance)/_t;
    v_Rr_a=v_Rl_a*(1+(b/l)*tan(delta_a))/(1-(b/l)*tan(delta_a));
    v_a=v_Rl_a/(1-(b/l)*tan(delta_a));
    write("\nv_a=");write(v_a);write(",
    v_Rl_a=");write(v_Rl_a);write(", v_Rr_a=");write(v_Rr_a);
    //——

    displayPar(delta_a,v_a);
    displayPosition(q1,q2,q3);

    //Calculating intermediate positions
    float t_int=_t/100;
    float x_int,y_int;
    q1_int[0]=q1;q2_int[0]=q2;q3_int[0]=q3;
    for(p=0;p<100;p++)
```

```
{
        //if global array used for whole problem then we've to use
        a global increment (ie p here).
        if(fabs(delta_a)>delta_a_min)
        {
                x_int=(1/tan(delta_a))*sin((v_a/1)*tan(delta_a)*t_int);
                y_int=(1/tan(delta_a))*(1-cos((v_a/
                1)*tan(delta_a)*t_int));
                q1_int[p+1]=q1_int[p]+(x_int*cos(q3_int[p])-
                y_int*sin(q3_int[p]));
                q2_int[p+1]=q2_int[p]+(x_int*sin(q3_int[p])+
                y_int*cos(q3_int[p]));
                q3_int[p+1]=q3_int[p]+((v_a/1)*tan(delta_a)*t_int);
                //not required
        }
        else
        {
                x_int=v_a*t_int;
                y_int=0;
                q1_int[p+1]=q1_int[p]+(x_int*cos(q3_int[p])-
                y_int*sin(q3_int[p]));
                q2_int[p+1]=q2_int[p]+(x_int*sin(q3_int[p])+
                y_int*cos(q3_int[p]));
                q3_int[p+1]=q3_int[p]+0;
                //not required
        }
}
/*for(int r=0;r<p;r++)
{
        write("\nq1_int[");write(r);write("]=");write
        (q1_int[r]);write(" ,q2_int[");write(r);
        write("]=");write(q2_int[r]);
}*/
//——

//Calculating final values of q1,q2 & q3
if(fabs(delta_a)>delta_a_min)
{
        x=(1/tan(delta_a))*sin((v_a/1)*tan(delta_a)*_t);
        y=(1/tan(delta_a))*(1-cos((v_a/1)*tan(delta_a)*_t));
        q1+=x*cos(q3)-y*sin(q3);
        q2+=x*sin(q3)+y*cos(q3);
        q3+=((v_a/1)*tan(delta_a)*_t);
}
```

```
else
{
        x=v_a*_t;
        y=0;
        q1+=x*cos(q3)-y*sin(q3);
        q2+=x*sin(q3)+y*cos(q3);
        q3+=0;
}
write("\nGlobal co-ordinates of vehicle CoG: [");
write(q1,4);write(q2,4);write(",");write(q3,4);write("]");
//——

//Calculating step timing array
if((((fabs(v_Rl_a)*_t)/step_distance)<1.002)||(((fabs(v_Rr_a)*_t)/
step_distance)<1.002))
{
        displayStatus("Parameters out of range. Exiting...");
        write("\nParameters out of range. Exiting...");
        closeGFX();
        exit(1);
        /*if(v_Rl_a<=v_Rr_a)
                v_Rl_a=(round(1.02)*step_distance)/_t;*/
}

unsigned int delay_Rl=(_t/(((fabs(v_Rl_a)*_t)/step_distance)-
1))*1000;
//delay in ms –>exit on delay=0 –>check for single step
unsigned int delay_Rr=(_t/(((fabs(v_Rr_a)*_t)/step_distance)-1))
*1000;
//int data type limits max value of delay to 65,534ms

unsigned int timer_Rl[1000],timer_Rr[1000],timer_main[2000][2];
//step timing array for Rl, Rr, and union of the two.
timer_Rl[0]=0;
//the second index in timer_main denotes the motor id to be stepped.
timer_Rr[0]=0;
int i=0,j=0;

while((delay_Rl*(i+1))<=(_t*1000))
{
        i++;
        timer_Rl[i]=delay_Rl*i;
}
while((delay_Rr*(j+1))<=(_t*1000))
```

```
        {
                j++;
                timer_Rr[j]=delay_Rr*j;
        }

        int k=1,l=1,m=0,flag;
        timer_main[m][0]=0; timer_main[m][1]=5;
        for(k=1;k<=i;k++)
        {
                flag=1;
                while((l<=j)&&(flag==1))
                {
                        if(timer_Rl[k]<timer_Rr[l])
                        {
                                timer_main[++m][0]=timer_Rl[k];
                                timer_main[m][1]=2;
                                flag=0;
                        }
                        else
                        {
                                if(timer_Rl[k]==timer_Rr[l])
                                {
                                        timer_main[++m][0]=timer_Rl[k];
                                        timer_main[m][1]=5;
                                        l++;
                                        flag=0;
                                }
                                else
                                {
                                        timer_main[++m][0]=timer_Rr[l];
                                        timer_main[m][1]=3;
                                        l++;
                                }
                        }
                }

                /*if(l>j)                              //still not tested
                {
                        timer_main[++m][0]=timer_Rl[k];
                        timer_main[m][1]=2;
                }*/
        }
        for(l;l<=j;l++)
        {
```

```cpp
                if(timer_Rl[i]!=timer_Rr[l])
                {
                        timer_main[++m][0]=timer_Rr[l];
                        timer_main[m][1]=3;
                }
        }

        if(stat_flag=='D'||stat_flag=='d'||stat_flag=='B'||stat_
        flag=='b')
        {
                cout<<"\nmotor Rl - "<<(i+1)<<" steps";
                cout<<"\nmotor Rr - "<<(j+1)<<" steps";
        }
        if(stat_flag=='L'||stat_flag=='l'||stat_flag=='B'||stat_
        flag=='b')
        {
                outfile<<"\nmotor Rl - "<<(i+1)<<" steps";/*cout<<" with
                step times:-\n";
                for(p=0;p<=i;p++)
                {
                        outfile<<setw(4)<<timer_Rl[p]<<"|";
                        if((p+1)%25==0)
                                outfile<<"\n";
                }*/
                outfile<<"\nmotor Rr - "<<(j+1)<<" steps";/*cout<<" with
                step times:-\n";
                for(p=0;p<=j;p++)
                {
                        outfile<<setw(4)<<timer_Rr[p]<<"|";
                        if((p+1)%25==0)
                                outfile<<"\n";
                }*/
                /*outfile<<"\ncombined step timing array:-\n";
                for(p=0;p<=m;p++)
                {
                        outfile<<setw(4)<<timer_main[p][0]<<"("<<timer_main
                        [p][1]<<")|";
                        if((p+1)%25==0)
                                outfile<<"\n";
                }*/
        }
        write("\n————————");
        //————
```

```
//Executing step sequence
displayStatus("Moving Vehicle");
step(2,dir);
step(3,dir);
p=0;
for(int n=1;n<=m;n++)
{
        delay(timer_main[n][0]-timer_main[n-1][0]);

        if(timer_main[n][1]==5)
        {
                step(2,dir);
                step(3,dir);
        }
        else
        {
                step(timer_main[n][1],dir);
        }

        while(((t_int*1000*p)>=timer_main[n-1][0])&&((t_int*1000*p)
        <=timer_main[n][0]))
        {
                plot(q1_int[p],q2_int[p],0);
                displayPosition(q1_int[p],q2_int[p],q3_int[p]);
                p++;
        }

}
//——
plot(q1,q2,0);
}

void main()
{
    const float pi=3.1416;
    float xd=-100, yd=100;
    // co-ordinates of destination point, values in mm
    float v0=20.0,v1=3.5,vpar=0.5,cpar=1; // parameters
    float t=2;                                  // time interval in s
    float e1,e2,e3;
    // local co-ordinates of destination point (mm,mm,radians)
    float x_err=0.5,y_err=0.5;
    // final closeness to destination point
    float rad,vel,angle,ang;
```

```cpp
char motion;

clrscr();
promptForStatus();
cout<<"\nDo you want a motion along an arc(c) or to destination
points(p)? ";
cin>> motion;
if(motion=='p')
{
        cout<<"\n";
        cout<<"\nEnter destination co-ordinates:-";
        cout<<"\nx=";
        cin>>xd;
        cout<<"y=";
        cin>>yd;
        initializeGFX(.3,.3);
        plot(xd,yd,1);
        plot(q1,q2,0);

        displayStatus("Execution Begins");
        while((fabs(xd-q1)>x_err)||(fabs(yd-q2)>y_err))
        {
                e1= (xd-q1)*cos(q3)+(yd-q2)*sin(q3);
                e2= -(xd-q1)*sin(q3)+(yd-q2)*cos(q3);
                e3= atan((yd-q2)/(xd-q1))-q3;
                delta_a=e3;
                if((abs(e1)>30)&&(e1!=0))
                {
                        v=v0*((e1)/abs(e1));
                        delta=e3;
                }
                else
                if(e1==0)
                {
                        v=v0;
                        delta=e3;
                }
                else
                {
                        v=v1*((e1)/abs(e1))+vpar*e1;
                        delta=0;
                        t=1;
                }
                moveVehicle(delta, v, t, -1);
```

```
            }
            displayStatus("Execution Complete");
            closeGFX();
    }
    else if(motion=='c')
    {
            cout<<"\n";
            cout<<"\nEnter the Details of the path:-";
            cout<<"\nradius=";
            cin>>rad;
            cout<<"velocity=";
            cin>>vel;
            cout<<"angle=";
            cin>>angle;
            initializeGFX(.3,.3);
            plot(xd,yd,1);
            plot(q1,q2,0);
            displayStatus("Execution Begins");
            ang=(pi/180)*angle;
            delta= atan(l/rad);
            v=vel;
            t=(rad*ang)/v;

            moveVehicle1(delta, v, t, -1);
            displayStatus("Execution Complete");
            closeGFX();
    }
}
```

APPENDIX B
ABOUT THE CD-ROM

Included on the CD-ROM are simulations, figures from the text, third party software, and other files related to topics in robotics.

See the "README" files on the CD-ROM for specific information/system requirements related to each file folder, but most files will run on Windows 2000 or higher and Linux.

INDEX